Why Einstein Was An Ignorant Fool

How Einstein, Being Ignorant of Today's Experimental Physics, Was Able to Fool the World into Believing in Many Impossible Things Like Infinite Fields, Massless Photons, Downward Pointing Gravity, Equivalent Momentum and Force and an Instantaneous Expanding Big Bang Singularity Instead of a Living Universe Duality.

James Carter

Absolute Motion Press
Franklin, Washington

Copyright © James Carter 2018
First Edition
All rights reserved. No part of this publication may be reproduced, stored in a retrieval system, or transmitted, in any form by any means, electronic, mechanical, photocopying, recording, or otherwise, without the prior permission of Absolute Motion Press. Anyone can quote from this book as long as James Carter is given credit.

Printed in the United States of America
September 11, 2018

James Carter
email: circlon@gmail.com

ISBN 978-0-359-13622-3
Black & white

Additional information about the Physical Principles of Absolute Motion and the Circlon Synchronicity of the Living Universe can be found at:
www.living-universe.com
www.circlon.com
www.circlon-theory.com

Other books available by James Carter

Physics without Metaphysics

Big Bang Physics Without Metaphysical Assumptions

Circlon Synchronicity in the Living Universe

These books can be purchased at www.Living-Universe.com
Or at www.LULU.COM

Introduction

If you have no knowledge of physics and chemistry, or don't understand the Standard Model Physics and Big Bang Theories, you will be able to easily understand the principles and believe the ideas presented in this book. Everything is explained in terms of the principles and logic of the scientific method of experimental measurement.

However, if you have multiple degrees in physics and cosmology, these basic ideas of mass, space, time, and gravity may seen to be too simple and naïve for your complex and critical mind to fully comprehend. The problem may be these ideas cannot be adapted to your metaphysical beliefs in a continuum of aethers, fields, and multiple spacetime dimensions that are all ultimately composed of "pure energy".

New beliefs in physics and cosmology will only come when you are able question the old beliefs acquired from your education. You could start by questioning your firm belief that the electron/proton mass ratio of 1/1836 is a universal and eternal constant that has not changed since the Big Bang.

After that, you can question your long held conscious belief that gravity points down and contrast it with your subconscious belief derived from your senses that gravity is felt as an upward push. You will never be able to convince your subconscious mind's sense of balance that Einstein's foolish little gravity theory could be true.

If you believe the cosmos was created by a god and goddess and not by a godless Big Bang singularity, then this book is for you. The Living Universe is divided between two eternally evolving entities. Its history is revealed in several fundamental stages from the time before the electron/proton ratio was 1/1 to the present day when it is 1/1836. You will learn that the Hubble shift is not what you believe it to be and you will want to understand the true nature of the 2.7°K Cosmic Blackbody Radiation and the real reason for its sudden appearance near the beginning of cosmic evolution. You can also learn why the dinosaurs were so big.

This book does not present any new "physics theories" for atoms, photons or gravity. No metaphysical assumptions are made about any of the scientific experiments described in this book. It is about how experimental physics is the subject of theoretical physics and not the other way around. The cosmos is what we measure it to be and not what we want or imagine it to be.

Einstein was an Ignorant Fool

This is how Einstein ignored experimental physics, imagined his relativity theories were true, and then went on to fool generations of theoretical physicists into believing in impossible things like infinite fields, massless photons, equivalent force and motion, non-local gravity, and the many metaphysical assumptions of the Big Bang creation myth.

Ignorance itself can be a virtue. We are born ignorant of all things and then try to learn all we can, but when we die, we are all still ignorant of many important things and ideas. What knowledge we choose to learn and believe determines the quality of our ignorance. While not a virtue, being a fool is an essential part of one's path to knowledge. Fools believe in things that are not true but that is a necessary part of their quest for truth. Fools use their imaginations rather than their senses to create their theories of the world.

Einstein said that, "Imagination is more important than knowledge". This is the creed of the theoretical physicist. It is opposite for the experimental physicist who believes knowledge to be the subject of imagination. Einstein imagined his theories by ignoring physical experiments and then used his equations to calculate alternative values for experimental measurements. His calculated values perfectly matched the measurements, but the parameters in his equations were the opposite of those being measured. As Einstein and his followers gained more and more knowledge of modern experimental physics, they were able to carefully fit their upside down foolish equations into the discovery of each new physical phenomenon. Einstein's theories contained perfectly correct calculations even though they were upside down, backwards and inside out from the experimental measurements of mass, space, time, and gravity.

Einstein's 3 Metaphysical Assumptions of Impossible Things

1. The rest energy/mass $e/m = c^2$ of matter can be converted into the pure momentum $p = mc$ and pure energy $e = mc^2$ of massless photons.

2. Gravitational force $g = ms/t^2$ and motion $V_{es} = \sqrt{2gr}$ are directed toward Earth's center.

3. The electron/proton mass ratio of $e/p = 1/1836$ is an eternal and universal constant.

Einstein's Postulates

To frame these three assumptions, Einstein felt it necessary to include two somewhat contradictory "postulates".

First postulate:
The laws of physics are the same in all inertial frames of reference.

This postulate is quite incomplete as a physics law because it is infinite and eternal. The postulate seems to imply that "the laws of physics" have always been the same. This idea is presented as a principle of measurement even though there is no experimental evidence to support its claims. It seems that this was just an idea that Einstein and all of his followers wanted to believe in but could not come up with any supporting evidence.

Einstein's First Postulate in Modern Physical Terms:

The laws of physics and cosmology based on the 1/1836 electron/proton mass constant are the same now, in the present, as they were in distant galaxies and all the way back to the beginning of the universe when the 1/1836 e/p mass ratio first came into existence with a bang. There is even speculation by some cosmologists that the 1/1836 e/p ratio may have existed as a law before the Big Bang.

Einstein foolishly painted himself into a corner with this postulate. Certainly, he considered the 1/1836 cosmic ratio to be one of the universal laws of physics. He knew that this ratio controlled the reciprocal values of the Bohr radius and the fine structure constant. However, he played the fool when he insisted that these constants were eternal. It forced him to conclude that today's constants and laws of electrodynamics are exactly the same as they were in the distant past. This led him to imagine, and then fool others into believing that the Hubble red shifts are Doppler effects caused by the increasing velocities of an expanding universe created from a big bang singularity explosion of "pure energy".

The alternative would be to ignore the eternal nature of the first postulate and imagine that the enormous Hubble shifts in spectral photons were caused by a slow evolution of the e/p mass ratio resulting in evolving physical laws of electrodynamics. The Hubble photons were emitted when the e/p ratio was smaller and the electrons within radiating atoms had more mass.

This idea eliminates the need for the vast momentum and energy resources of the Big Bang creation and allows for the conservation of momentum and energy/mass. In Big Bang theory, there are no conservation laws for momentum and energy. It is assumed that more than 99% of the original momentum and energy created in the singularity has mysteriously disappeared into some unknown dimensional void.

What did Einstein mean by "laws of physics"? It would seem that the conservation laws would be considered laws of physics but they are clearly violated by his first postulate.

Second postulate: Absolute Speed of Photons c & C
As measured in any inertial frame of reference, light is always propagated in empty space with a definite velocity c that is independent of the state of motion of the emitting body.
Or: The speed of light in free space has the same value c in all inertial frames of reference.

This postulate is quite incomplete because Einstein does not make a distinction between one-way and two-way measurements of c. Photons are always emitted at c relative to the same zero momentum inertial frame but one way measurements are never at c due to an observer's unknown absolute momentum. Two way measurements are always at c. The postulate also fails to consider the photon's rotational speed of light C and its relationship to the rotational velocity C of the emitting and absorbing atoms.

2nd Postulate in Modern Physical Terms:
Photons always move through empty space at the linear speed of light c and always rotate in empty space at the rotational speed of light C. Photon motion is measured as momentum $p = mc$ and angular momentum $I\omega = m\lambda C/2\pi$. The photon's energy/mass is equal to the two separate speeds of light $e/m = cC$. A photon's measured energy $e = mcC$ is relative to the observer's frame for its linear energy $e = mv^2/2$ and constant in all frames for its rotational energy $e = mC^2/2$.

The second postulate is not an assumption, it is a principle of measurement that is always verified and quantified by experiment. It is a basic law of experimental physics that all photons move at c & C within the same inertial frame of reference.

Einstein's 2nd postulate was about measuring the speed of light. However he was fooled by the measurement process in which two-way photon measurements are always at c but one-way measurements of photon speed always include the relative velocity of the observer. Photons never move at c relative to any moving observer's reference frame and are always measured at c+/-v.

The Lorentz Transformation $\sqrt{1-v^2/c^2}$

In addition to his three metaphysical assumptions and two postulates, Einstein adapted the Lorentz Transformation $\sqrt{1-v^2/c^2}$ of mass and time to both of his theories. The Lorentz Transformation is a physical principle of measurement and not a metaphysical theory because it is easily calculated and quantified whenever motion is measured. It is a physical principle, that is used in almost all physical theories for the measurement of absolute motion, force, and energy. All three of Einstein's theoretical assumptions are metaphysical in nature because they cannot be independently verified or quantified by direct physical measurement.

Einstein's False Conclusions from His Assumptions and Postulates

* *Massless photons travel through an electromagnetic field medium called "spacetime". This allows "pure energy" photons to move through space at c while "pure mass" atoms remain at rest.*

* *Gravitational force and motion are equivalent but not equal to inertial motion and force. This causes the direction of gravity's motion and force to point down toward Earth's center.*

* *The Hubble red shift is a relative velocity Doppler effect. The direct cause of the "shift" is the eternal constant of the electron/proton mass ratio 1/1836.*

Einstein's Foolish Mistakes

With these three unmeasurable ideas, Einstein was able to first fool himself and then went on to fool most Twentieth Century theoretical physicists into imagining and then believing in many impossible things that could be calculated but not measured by experimental physicists. This unmeasured metaphysical foolishness culminated in the Big Bag theory of creation where common sense and all of the well established fundamental laws of physics have to be abandoned at one time or another.

Massless Photons

Measurements of photon momentum p = mc, angular momentum Iω = mλC/2π, and energy e = mc²/2 + mC²/2 = mcC all quantify photon mass. No photon experiment has ever measured zero mass or even intimated that a zero mass photon could exist. It can be said that Einstein was the first to imagine and then invent the concept of the photon. Out of convention with earlier aether theories of light, he foolishly imagined photons to be massless waves of "pure energy" moving through an electromagnetic field medium at the constant speed of c. Although Einstein originally claimed not to believe in "aether", he later admitted that his proposed 4-dimensional spacetime continuum was indeed an aether-like medium.

Equivalent Force and Motion

All measurements with accelerometers and clocks show that gravitational force and motion are directed away from Earth's center and that they are equal and not just equivalent to inertial force and motion. Einstein had no experimental justification for adopting his counter-intuitive equivalence principle that predicts the opposite of what we measure.

1/1836 Electron/Proton Mass Ratio Constant

All cosmological evidence shows that the Hubble shift, Dark Energy, nuclear synthesis, and the temperature of the 2.7°K Cosmic Blackbody Radiation are all the result of a gradual increase in the electron/proton mass ratio of 1/1836. No evidence has ever been offered to suggest that this ratio and Einstein's other "laws of physics" are universal constants that do not change over time.

E = MC² is Wrong and E/M = cC & E/M = CC are Right

Energy and mass are two sides of the same coin and are always equal and cannot be physically separated in any conceptual way. e/m = cC is the formula for photons and e/m = CC is the formula for atoms. A moving body's value for energy/mass is e/m = v² - c²/c. At zero momentum rest, energy/mass equals zero, e/m = 0² - c²/c = c/0. At the speed of light, energy/mass would become infinite e/m = c² - c²/c = 0/c.

Einstein Wrong?

This does not mean that calculations of physical events made with Einstein's reversed equations of cause and effect are "wrong" because they always yield the correct values. The basic problem with Einstein's relativity theories is that they are completely unnecessary. He makes a theory that is the opposite of the real thing and imagines that his opposite direction of time is equivalent to the direction of time in Newtonian attraction. The physical difference between the attraction theories of Newton and Einstein is that with Einstein's equivalence, the direction of time in the cause and effect of gravitational interactions is reversed.

Einstein's relativistic calculations of massless photons come out exactly the same as the absolute motion of photons with mass. The physical photon equation is $e/m = cC$. Einstein's massless photon equation $e = mc^2$ is the same equation with the components of the photon separated from one another with a reversed direction of time for cause and effect events.

The calculations of gravity's measured upward force and motion yield identical results to Einstein's predicted values of equivalent and relative downward force and motion. Einstein's relative motion calculations always come out the same as actual absolute motion measurements. Einstein's calculations aren't "wrong" because his results are always "correct".

The Logic of the Living Universe versus the Magic of the Big Bang
Where Einstein's equations, calculations, and predictions really fall into the realm of actually being wrong is when he and his followers try to use his relativity theories to describe the elements and progression of the sudden Big Bang creation of atoms, stars, and galaxies.

The common idea of the Big Bang imagines the cosmos beginning as a magic singularity of pure energy that suddenly appeared in the center of the universe and exploded outward. Since that beginning of time, the initial pure energy of the singularity has divided and transformed into protons, electrons, and photons that eventually spread out into the cosmos to form galaxies, stars, planets, and people. This is the primary metaphysical assumption of all big bang theories and except for people there is no physical evidence for how, why or when any of this happened. Each big bang theorist has his own way of weaving metaphysical ideas together in a way that does not violate too many natural laws and physical principles of measurement.

In big bang theories, the singularity is an effect without a cause. The big bang is said to have "exploded" from a point into electrons and protons with an eternal mass ratio of $e/p = 1/1836$. Then almost instantly, they traveled to the far reaches of the universe on a magic carpet called Guth inflation. Once there, they began to cool and couple into neutrons, Hydrogen and other atoms. These widely spaced atoms then somehow managed to condense into clouds, stars, and galaxies. This description is based on the purest of imaginary speculation and except for the existence of galaxies there is no physical evidence for any of it. Most of the elements in the big bang creation stories represent effects without causes.

Theoretical physicists are allowed to imagine and calculate how, when, and why this pure energy spacetime field came into being at the beginning of forever and then transformed itself into the atoms, stars, and galaxies of today.

Theorists imagine eternal laws like the 1/1836 e/p ratio and then combine them with other imagined laws of physics, that do not exist today like inflation and laws that can magically convert pure energy into atoms and then instantly move them throughout the universe without conserving momentum. Experimental physicists have never been able to transform pure energy into atoms nor instantly transport them to distant locations,

Living Universe Logic

The logic of the Living Universe is a cosmos made out of electrons and protons that together produce and absorb photons. This is true today and both has been and will be true forever. If you really want to know the origin of the electron and proton, it can only be said that they are an eternal god and goddess duality that have lived forever. The early history of the Living Universe is explained in terms of an evolving electron/proton pair with a mass ratio that was once e/p = 1/1 and then slowly grew to its present day value of e/p = 1/1836.

There is no logical or experimental leap of faith in the cosmic evolution of the Living Universe because it begins with a positron and antiproton and ends with equal numbers of electrons and protons. Certainly, today everyone knows that the universe appears to contain equal numbers of electrons and protons. Most theorists make this fundamental assumption at the beginning of their theories and then try different methods and equations to explain why this is true.

The Living Universe is a principle and not a theory. Unlike relativity theories, there are no violations of Newton's laws, quantum mechanics, electrodynamics, or the laws of gravitational force and motion. Scientific instruments are used to measure electrons and protons but ultimately the instruments are all made out of electrons and protons. Electrons, protons, and photons are at the foundation of every measurement we can make of mass, space, time and gravity.

The Living Universe is a complete description of the cosmic evolution of electrons, protons, and photons but it is not a "theory" of the electron and proton. It is just their measurements. The electrons and protons in the Living Universe are the same electrons and protons of everyday measurement. Whatever way theorists picture the electron and proton in their minds or calculations, their conceptual models will need to fit the experimental measurements.

Einstein was a Brilliant Theoretical Physicist

The title of this book is not meant to be a meaningful insult of Einstein or his brilliant relativity theories. To say that he was ignorant is more of a complement than an insult. Even his staunchest supporters will tell you he was ignorant. Even Einstein himself claimed ignorance when he said, "Imagination is more important than knowledge."

What made his theories particularly remarkable is that at the time he created them, he was completely ignorant of the many aspects and phenomena of modern physics and astronomy that is common knowledge to all of today's physics students. When he presented the theory of Special Relativity in 1905 and invented a concept of the photon, he knew nothing of today's many sophisticated photon measurements. He didn't know about positrons, antiprotons, neutrons or neutrinos. He didn't know about the Compton effect, the Sagnac effect, or lasers and even claimed ignorance of the Michaelson-Morley experiment.

When he published the theory of General Relativity in 1915, he was unaware of galaxies, the Hubble red shift, Dark Energy, Dark Matter, Quasars, Pulsars, or neutrons stars. At the time of his death in 1955, the Pound-Rebka experiment had yet to be performed, the 2.7°K Cosmic Blackbody Radiation had not been discovered and there was no need for anyone to think about GPS satellite clock adjustments. The true genius of Einstein and his theories was that as each of these new phenomena came to light, he or his followers were able to incorporate the new data into his theories in precise if somewhat convoluted ways.

The Hubble Constant

When he first learned about Hubble's red shifted photons from distant galaxies, Einstein appears to have given them little thought before amending his General Relativity theory to remove the cosmological constant and allow for an expanding universe caused by a Big Bang singularity. In this, he was trying to agree with the ideas of most cosmologists even though Edwin Hubble's doubts remained as to the red shift's true cause. Hubble, who was an experimental physicist, gave the red shifted galaxies much more thought than Einstein. He wanted to take the shifts at face value and explain them in terms of existing physics rather than adopt new metaphysical assumptions.

Einstein's conclusion that the Hubble Shift was a Doppler effect led step by step to such imaginary metaphysical concepts as an expanding universe that emerged as pure energy from a spacetime singularity. Through unknown physical interactions, the pure energy of the singularity condensed into electrons, protons, and photons that were spread to the far reaches of the universe by the immaculate miracle of Guth Inflation. When these particles combined into atoms, they radiated a great burst of photon energy that is still observed today as the 2.7°K Cosmic Blackbody Radiation.

Einstein's theory, with its constant 1/1836 electron/proton mass ratio, predicts that this radiation would have been about 3000°K when emitted and has since cooled to 2.7°K. To imagine how this might have happened, Einstein's apologists designed a multi-dimensional spacetime continuum that can be described as an expanding photon space. This new type of aether slowly decreases the momentum and energy of all photons but has no effect on the momentum and energy of atoms moving through it.

By clinging to their e/p = 1/1836 assumption, Einstein enthusiasts believe more than 99% of the momentum and energy produced in the early universe has since dissolved into the fabric of their new spacetime aether. They go so far as to say this peculiar dimension of expanding space, is the ultimate reason behind our very existence. Were it not for this miraculous expanding space solution to Olbers' paradox, the Cosmic Blackbody Radiation would still be at 3000°K and we would all burn up!

The fact that Einstein carefully played the fool as well as the fooled is at the heart of his fascinating story. He claimed to his many followers that when we measure massless photons, we are really detecting the electromagnetic field that they travel through. He fooled them into believing that photons are emitted from and absorbed into preexisting electromagnetic fields and not the atoms themselves. This is in deference to experimental physicists who measure photons being emitted and absorbed by atoms directly.

In the force and acceleration of Newtonian physical measurement, photons have a momentum of $p = mc$, an angular momentum of $I\omega = m\lambda C/2\pi$, and an energy of $e = mcC$. Mass is common to the photon in all of these measured parameters. The reason Einstein imagined a massless photon was to invoke the magical properties of his idea for a four-dimensional spacetime continuum field. Without mass, a photon would need a medium like an aether, field, or spacetime continuum to carry its momentum and energy across the universe. A photon with mass needs no medium and its own Newtonian inertia can carry its energy, momentum, and angular momentum through empty space. While a photon's mass cannot be measured directly, it is inferred by a photon's momentum and energy that can be measured.

Photons e/m = cC & Atoms e/m = CC

Photons are measured to be particles of matter that travel back and forth between atoms where they can be reflected, absorbed, or emitted. They move through space like rifle bullets. They move and spin through space at the speed of light. They are measured to have a linear energy of $e = mc^2/2$ plus a rotational energy of $e = mC^2/2$. This is in contrast to the ambiguous energy of Einstein's massless photon $e = mc^2$ where there is no distinction made between linear and rotational momentum and energy.

The energy/mass in the photon equation is a unit that is equal to the linear times the rotational speed of light. $E//m = cC$. In this equation, the variable for Doppler effects is relative c. The energy/mass equation for matter at rest e/m = CC has the same basic value as the photon except that the mass within atoms is spinning at rotational C on two opposite planes with no linear motion relative to c. Whereas a photon has equal quantities momentum and angular momentum, an atom at rest has only two equal quantities of angular momentum until it is accelerated,

While Einstein didn't believe in the *energy/mass* of photons he did believe in the *energy/mass* of atoms. He fooled himself into believing that this conserved constant unit could be separated into massless photons with momentum and wavelengths but no mass. Einstein's massless photons have equivalent momentum and angular momentum of $p = mc$ and $I\omega = m\lambda C/2\pi$.

There is no experimental way that a massless photon can be detected except by assuming it has momentum equivalent to the mass it was created from and then measuring this equivalent momentum just like it was real momentum to determine the massless photon's wavelength $\lambda = h/mc$. Einstein was never able to quantify his massless photons except by using the parameter of mass at every step in his calculations.

Einstein was a Gravitational Magician

Einstein's greatest magical trick was to use smoke and mirrors to reverse the measured direction of the force and motion of gravity. His second assumption that ***Gravitational force*** $g = ms/t^2$ ***and motion*** $V_{es} = \sqrt{2gr}$ ***is directed toward Earth's center*** is only equivalent and opposite to gravity's true measured direction away from Earth's center.

In order to assume an equivalent and metaphysical down direction for gravity that could not be measured, it was necessary for Einstein to imagine and quantify a universe wide continuum field to contain all of the gravitational forces and motions within the Cosmos. Einstein imagined that through his magical continuum of gravitational fields, waves, and "curving" spacetime, each atom in the universe was intimately and eternally connected to every other atom.

The only way for an experimental physicist to understand the physical direction of gravity is to measure the dynamic of gravitational force and motion at the surface of Earth and then extrapolate the values to satellites, planets, stars and galaxies.

If Einstein had accepted local gravitational measurements at face value, there would have been no need for him to imagine and construct his dubious four-dimensional spacetime continuum. If gravity can be explained completely as locally produced force and motion, why would anyone want to imagine a metaphysical force extending to the far reaches of the universe when no measurement or calculation requires it.

Einstein's foolish blunders were his interpretation of the cosmological constant and the equivalence principle of gravity and inertia. The equivalence principle is the purest of metaphysical fantasies and the cosmological constant is caused by the slow evolutionary increase in the electron/proton 1/1836 mass ratio.

If Einstein was really as clever as everyone said he was, why didn't he at least try the simple math for the measurements of the gravitational expansion of mass, space, and time and then follow the calculations until he arrived at an impossible solution. Perhaps, he did but then realized that gravitational expansion was such a simple and counter intuitive concept that he wouldn't be able to fool anyone into believing it except for infants and small children who already believed in it. Einstein must have believed the gravitation expansion of mass, space, time to be impossible simply because almost everyone already believed in Newton's attraction theory between bodies of mass. Einstein simply modified this idea with the addition of his spacetime continuum field to provide the physical means by which the attraction could occur.

Einstein's $E = MC^2$ is Wrong and $E/M = cC$ and $E/M = CC$ is Right

The formula $e = mc^2$ was developed by several physicists before Einstein but when he adopted it as $m = e/c^2$, he imagined it to have new meanings. He combined it with the Lorentz transformation so when a body's energy is increased, its mass is also increased by a proportionate amount so that energy/mass maintains a constant ratio. When an atom is accelerated, its energy/mass is increased and when it is decelerated its energy/mass decreases. Einstein then went on to falsely conclude that when matter and antimatter combine they convert their rest mass into pure energy photons with momentum but no mass. This, of course, is one of the impossible things that all of Einstein's followers must attempt to believe in. How can you have momentum without mass?

From a basic experimental point of view, Einstein's elaborate spacetime continuum is superfluous and not at all necessary to explain and quantify Earth's upward gravitational force and motion. Gravity is just the complementary force and motion that we feel and measure and does not require any metaphysical assumptions or theories to fully explain how it works. Gravity is measured as a combination of acceleration and deceleration in the gravitational expansion of mass, space, and time. Gravitational expansion may be difficult for rigid minds to believe in but logical minds can easily quantify and calculate its local parameters.

Einstein's Foolish Gravity

Einstein took the simple local mechanical upward push of gravity that everyone constantly feels from infancy and tried to imagine it as a metaphysical pulling force spanning the entire universe. Einstein was a fool to disregard the feelings from his own sense of balance to imagine a downward pulling force that couldn't be felt. Einstein invented equations to calculate gravitational force and motion that were the mirror images of actual gravity. Einstein simply didn't understand the dynamic difference between up and down.

Einstein's Foolish Photons

There is nothing in photon measurements that would lead one to believe that they exist only as parts of an underlying spacetime medium field. A photon is measured as an absolute unit that is not connected to or part of anything else. Photons are measured as individual quantities of energy/mass $e/m = cC$ that move through one-dimensional space at the speed of light c as they spin in two-dimensional space at the rotational speed of light C. Einstein thought he was clever by separating energy/mass in this standard photon equation. His new equations, $e = mc^2$ & $m = e/c^2$, made it possible to conceptually separate energy and mass and imagine a photon without mass and an atom without internal energy.

The equations for photons and atoms are basically the same. In the equation for the energy/mass of atoms, protons and electrons, mass is at linear rest and spinning at the speed of light on opposite planes $e/m = CC$. The photon's energy/mass $e/m = cC$ moves at both linear c and rotational C. Einstein accomplished nothing but confusion by trying to bifurcate the energy/mass conserved constant into two equal and separate parts.

Einstein's Foolish Creation Story

Had Einstein allowed the electron/proton mass ratio to grow over time, he could have used his otherwise workable theories to describe a far more credible creation of matter and energy than the standard model Big Bang theories.

Despite his metaphysical assumptions, Einstein's relativity equations work very well to calculate accurate results for experimental measurements of all sorts of natural phenomena. Where his relativity theories and equations quickly fall apart is when they are used to predict and calculate the workings of the early universe. His foolish adoption of the 1/1836 electron/proton mass ratio as a universal and eternal constant led him imagine and then to believe in an exploding Big Bang creation instead of a Living Universe with an evolving electron/proton mass ratio.

To establish a credible beginning for the substance of reality that everyone can follow and understand, we must begin with the physics-neutral entities of gods. Einstein once indicated that he was trying to discern the mind of God. The principles of physics are not just ideas in the minds and thoughts of gods. Physics is actually contained in the bodies and souls of the gods. The electron and proton gods did not "create the universe", they are the cosmos. Einstein's silly idea was to try and make a universe containing something from a pre-universe of nothing. This is backward thinking that ignores the experimental principles of cause and effect. All measurements show that the cosmos is going from something toward nothing: entropy. Cosmic history cannot go from nothing to something when we can see it going from something to nothing. This is a cause and effect direction of time that cannot start at nothing. Matter can spread out into virtual nothingness but it cannot come from nothingness.

The Creation and Evolution of the Living Universe

The energy that produced the Big Bang occurred from the standard atomic and nuclear physics that has been adopted by the various interpretations of quantum mechanics. There is no physical evidence to support the Big Bang idea that the universe began with the sudden appearance of a "Pure Energy Spacetime Singularity". The Living Universe began with a simple eternal duality containing the same amount of conserved energy and mass contained in the cosmos today. This matter/antimatter duality divided into the equal numbers of protons and electrons contained in the cosmos today. The process by which this primordial duality was transformed into today's atoms and photons is in complete compliance with Newton's laws of motion, nuclear physics, and the quantum mechanical laws of electrodynamics. These laws and calculations are applied to the evolving 1/1836 electron/proton mass ratio and the effect that its changing value has on atomic constants like the Bohr Radius, the Fine Structure Constant and the Hydrogen Ionization Photon.

The only difference between the principle of the Living Universe and the Big Bang theory is in the way the original mass and energy of the universe was divided up into individual atoms and photons and spread throughout the cosmos. In the Living Universe, this process follows the well established laws of motion and quantum mechanics. In the Big Bang, the Newtonian laws of motion and quantum mechanics are mostly ignored and the ideas for dividing up atoms and photons and spreading them throughout the universe are created with several different smoke and mirrors mathematical inventions of various spacetime dimensions, fields, and aethers that all violate the basic laws of quantum mechanics.

Big Bang Theories versus Living Universe Principles

The Living Universe is based on the scientific method of measurement and makes none of Big Bang theory's unverifiable metaphysical assumptions. The initial physical assumption of the Living Universe model of creation is based on dual experimental principles.

Electrons and protons are eternal and always exist in equal numbers.
&
Photons are eternal energy/mass dualities created from the interactions of electrons and protons.

These dual principles were true when the cosmos began, are true today, and will be true in the future. These are not metaphysical assumptions because they are verified by all experimental measurements. The Living Universe model contains no metaphysical assumptions and is based completely on the validity of these two principles of experimental measurement.

In contrast, the Big Bang theory is constructed around a series of often contradictory metaphysical assumptions. Big Bang's initial violation of natural law is the creation of matter without antimatter. In the Living Universe the 1/1 ratio between matter and antimatter particles has always remained constant. This simple law of quantum mechanics and not a theory. The electron was once the antiparticle to the proton.

Einstein's Big Bang universe began with an enormous amount of "pure" energy at a virtually infinite temperature that has been imagined to be cooling off and disappearing ever since. The Living Universe began with a conserved amount of energy/mass at a temperature of absolute zero $0°K$ and has been gradually warming up ever since, to today's temperature of about $6°K$.

Big Bang theories violate the conservation of momentum and the first and second laws of thermodynamics because they have no fixed relationship between energy and mass as in their equation $e = mc^2$. Relativists believe the universe began as all energy and that it now contains mostly mass. In the Living Universe energy and mass have always been equal and are actually the same thing. Energy/mass are equal and opposite aspects of atoms and photons $e/m = CC$ & $e/m = cC$

Big Bang enthusiasts imagine the mass contained in electrons and protons can be transformed into massless pure energy photons. They assume photons throughout the universe constantly lose momentum and energy that disappear without a trace into the fabric of one of their several different kinds of expanding space. In the Living Universe, there is absolute conservation of a photon's energy/mass, momentum, and angular momentum and there is no "space" other than the negative reality of an imaginary universal void. This law requires the eternal existence and equality of a photon's energy/mass, $c/m = cC$. When a photon is absorbed by an atom, it maintains its energy/mass until the atom emits it.

Big Bang theorists illustrate their ideas with a multitude of complex mathematical equations. The Living Universe equations listed below simply illustrate the mass, space and, time relationships between electrons, protons, neutrons, and photons as calculated with the parameters of this new circlon shape interpretation of standard model quantum mechanics.

Living Universe Equations

Electron/Proton Mass Ratio Today --- $m_e/m_p = 1/1836$
Bohr radius --- $a_o = \lambda_\infty \alpha/4\pi$
Electron Angular Momentum --- $I\omega = m_e a_o \alpha c$
Fine Structure Constant --- $\alpha = 4\pi a_o/\lambda_\infty$
Photon Energy/Mass --- $e/m = cC$
Hydrogen Ionization Photon --- $\lambda_\infty = 4\pi a_o/\alpha$
Photon Momentum --- $p = mc$
Photon Angular Momentum --- $I\omega = m\lambda c/2\pi$
Electron Momentum --- $p = m_e v$
Electron Angular Momentum --- $I\omega = m_e a_o \alpha c$
Electron kinetic energy --- $e = m_e v^2/2$
Neutron Stability Number --- $\#_{ns} = m_p/m_e \sqrt{\alpha}$

The Initial Condition of the Living Universe

Einstein believed the initial condition of the universe was a gigantic explosion of pure spacetime energy that mysteriously transformed into photons, hydrogen and other atoms and then instantly spread to the far reaches of the universe. As usual, Einstein got it backwards. The Living Universe began as a gigantic atom of pure energy/mass that synchronously evolved into the photons, neutrons, hydrogen, and the other atoms in today's cosmos. The cosmos began as just a single antiproton/positron pair coupled into an anti-hydrogen atom sitting at zero momentum rest in the centers of an infinite number of one-dimensional voids.

All of the conserved energy/mass, momentum, angular momentum and charge energy of today's universe was contained within this single primordial ground state atom. The energy/mass of atoms is a universal constant $e/m = CC$. The energy/mass of today's cosmos is exactly the same as the energy/mass of the original anti-Hydrogen atom.

This initial state of zero momentum atomic rest defines the anti-atom's beginning temperature of absolute zero $0°K$. The atom has the rotational energy of its mass spinning in four different directions at the rotational speed of light C, It has angular momentum but no linear momentum to create the linear kinetic energy that defines temperature.

The Living Universe began with a standard model theory ***positron*** of constant and eternal mass and a standard model ***antiproton*** with slowly decreasing mass. The rate of antiproton and then electron mass decrease is the true physical manifestation of time in the universe. It is this constant change in the values and properties of electrons that makes the atoms and neutrons of the past dynamically different from the atoms and neutrons today. This is the e/p mass ratio clock with today's Electron Evolution interval of time at $T_{EE} = 1/1836$.

The only constant in the cosmos is the structural relationship between mass, space, time, and gravity and their calculations of force, momentum, angular momentum, and energy. The history of the Living Universe is represented below as time points and intervals of the e/p ratio clock. It goes from the evolution of the antiproton/positron mass ratio beginning at a/p = 1836/1 and ending today at an electron/proton mass ratio of e/p = 1/1836. This is a non-steady state universe that is eternally evolving into a new and different future based on the values and equations listed above.

Nine Stages in the Cosmic Evolution of Electron Mass

The history of the universe is represented below in nine separate intervals in the evolving electron/proton mass ratio. The Living Universe has steadily evolved from the time when the negative/positive mass ratio of atoms of matter was antiproton/positron = 1836/1 until today when the electron/proton mass ratio is e/p = 1/1836. In the following equations, I use the symbol (c) for the relative linear speed of light and the symbol (C) for the absolute rotational speed of light so that the energy of the atom is e = mCC and the energy of the photon is e = mcC.

An Arbitrary Beginning
a/p mass ratio 1836/1
Bohr radius $a_o = \alpha \lambda_\infty/4\pi = 1 = 5.292 \times 10^{-11}$
Fine structure constant $\alpha = 4\pi a_o/\lambda_\infty = .0073$
Hydrogen ionization photon $= \lambda_\infty = 4\pi a_o/\alpha = 9.11 \times 10^{-8}$
Neutron Stability Number $= M_P/M_E \sqrt{\alpha} = 157$

A single eternal antiproton/positron atom already existed in the universe at the "beginning" of the cosmos. These two equal and opposite eternal energy/mass particles were coupled together by their mutual charge chains into a ground state anti-hydrogen atom. All of the universe's energy/mass e/m = CC was contained completely within this atom as rotational energy/mass with no linear energy/mass.

The Antineutron Era
a/p mass ratio 146/1
Bohr radius $a_o = \alpha\lambda_\infty/4\pi = 12.5 = 6.615 \times 10^{-10}$
Fine structure constant $\alpha = 4\pi a_o/\lambda_\infty = .0000465$
Hydrogen ionization photon $= \lambda_\infty = 4\pi a_o/\alpha = .000179$
Neutron Stability Number $= M_p/M_E\sqrt{\alpha} = 1$

After a long period of time, the antiproton mass decrease and size increase reached the point where the neutron stability number goes from completely stable at less than one to slightly unstable at one. This allowed the positron to be captured by the antiproton and create an antineutron without the need of additional energy. The Living Universe was now a single antineutron sitting at zero momentum rest and about to decay.

The God Particles of Creation
a/p mass ratio 147/1 to 1/1
Bohr radius $a_o = \alpha\lambda_\infty/4\pi = \infty$
Fine structure constant $\alpha = 4\pi a_o/\lambda_\infty = 1/\infty$
Hydrogen ionization photon $= \lambda_\infty = 4\pi a_o/\alpha = 4\pi$
Neutron Stability Number at $1/1 = M_p/M_E\sqrt{\alpha} = 0$

As this primordial antineutron sat at the center of the universe, with the positron spinning inside of its internal circlon coil structures, the antiproton continued to decrease in mass. In order to conserve the particle's angular momentum, the antineutron's internal linear energy/mass increased in proportion to the rotational energy/mass lost. This increasing energy eventually made the antineutron unstable and it bifurcated into a pair of antineutrons that were ejected out into a one-dimensional void at somewhat less than the speed of light.

All of the linear energy/mass that had accumulated in the original antineutron was contained in the opposite velocities of the two new antineutrons and they contained only rotational energy/mass. Each particle then slowly began to again accumulate linear energy/mass as the antiproton gradually decreased in mass.

Then after a very long time, perhaps millions of years, these two god particle gained too much linear energy/mass to remain stable and they simultaneously bifurcated into four antineutrons. Then, after another long incubation period, these four particles bifurcated into eight and then, much later, into 2^4. Each time, all the particle's accumulated energy was used to propel them in opposite directions and each new antineutron was at virtual rest in its new moving frame.

The Living Universe now contained sixteen god particles millions of light years apart and rapidly moving outward in all directions. This began a long slow process of serial bifurcation that went through about 256 cycles in perfect synchronicity to eventually create 2^{256} antineutrons.

This was a non-linear process in which the time intervals between each cycle grew progressively shorter and the amount of kinetic energy between the ejected particles became less and less. The result of this is that the antineutrons from the earlier bifurcations had very high velocities and at the same time had long periods of time to spread out into the void of space. Then as the cycles speeded up and the neutrons has less energy, they began bifurcating faster, closer together, and with less velocity. At the end of the process, the bifurcations became almost instantaneous before stopping at e/p ratio 1/1+ where the electron was less massive than the proton and the very unstable antineutrons became 2^{256} stable neutrons.

The end result of this progressive process was to first spread energy/mass particles out into the far reaches of the universe and then toward the end, when the bifurcations occurred much faster and closer together, the bifurcating antineutrons were concentrated in vast individual clouds of virtually stable neutrons that would eventually decay and become stars and galaxies.

The bifurcation process ended when the antiproton/positron mass ratio became 1/1. At this point, the particles became electron/positron pairs that bifurcated one last time into pairs of stable neutrons. Once the e/p became less that one 1/1+, the antiproton was conceptually transformed into an electron that was less massive and larger than the proton. As the electron lost mass as it spun inside of the neutron, it gained kinetic energy in proportion to its mass decrease. In the case of the neutron, its increasing energy from mass loss does not make the particle unstable because the negative particle is inside of the positive particle and getting bigger. In the antineutron the negative particle is on the outside of the positive particle and getting bigger. As the e/p ratio moves forward, antineutrons become less stable and neutrons become more stable.

This long multi-billion year process by which the god particle antineutrons slowly bifurcated and spread out into the cosmos replaces the Big Bang ideas involving the so called Guth inflation where the whole expansion of the universe occurred virtually instantaneously. Guth inflation was proposed long after Einstein but it is a foolish ad hoc metaphysical idea that has no relationship at all to any of the standard laws of physics or any scientific measurements.

Toward the end of this long process of bifurcating god particles, the antineutrons became so numerous that they occasionally collided with one another. This unbalanced their internal energy levels and caused the antimatter/matter pairs within them to immediately annihilate into a pair of gamma photons instead of bifurcating into neutrons. By the time all of the god particles had been transformed into stable neutrons, the universe then contained a large number of very high energy gamma photons moving in all different directions. Today, we still observe these photons as the highly energetic cosmic rays.

The Neutron Cloud Bomb
e/p mass ratio 1/146
Bohr radius $a_0 = \alpha\lambda_\infty/4\pi = 12.5 = 6.615 \times 10^{-10}$ m
Fine structure constant $\alpha = 4\pi a_0/\lambda_\infty = .0000465$
Hydrogen ionization photon $= \lambda_\infty = 4\pi a_0/\alpha = .000179$ m
Neutron Stability Number $= M_P/M_E \sqrt{\alpha} = 1-$

At this point in history, the electron/proton mass ratio has reached 1/146.5 and the Neutron Stability Number passed from stable at a value of less than one to unstable at greater than one. This caused all neutrons throughout the universe to decay with great synchronicity into rapidly moving electrons and protons. These 2^{257} high energy electrons and protons either collided and recoiled from other particles or coupled with them to form neutrons, alpha particles, nuclear isotopes, atoms, and chemical compounds. In this process, nearly half of the universe's angular momentum and rotational energy was converted into the linear momentum and kinetic energy of 2^{257} rapidly moving electrons and protons.

The Universal 2.7°K Grand Fire
e/p mass ratio 1/147
Bohr radius $a_0 = \alpha\lambda_\infty/4\pi = 12.49 = 6.61 \times 10^{-10}$ m
Fine structure constant $\alpha = 4\pi a_0/\lambda_\infty = .0000468$
Hydrogen ionization photon $= \lambda_\infty = 4\pi a_0/\alpha = .000177$ m
Neutron Stability Number $= M_P/M_E \sqrt{\alpha} = 1+$

At this stage of cosmic evolution, the universe consisted of rapidly moving electrons, protons, alpha particles, and other nuclear isotopes and atoms. The electrons began coupling with atomic nuclei to form atoms. In this process, the kinetic energy of their motions and the electron/proton ionization charge energy were transformed into the linear energy/mass of photons e/m = cC. At an electron/proton mass ratio of 1/147, atoms emit blackbody spectral photons at a temperature of about 2.7°Kelvin.

Eventually, the atoms emitted their last thermal photon and dropped down into ground states. The universe was now filled with mostly ground state atoms that had converted equal parts of their momentum into the momentum and energy of photons. These photons traveled the universe with constant conserved momentum p = mc, angular momentum $I\omega = m\lambda c/2\pi$, energy e = mcC, and wavelengths λ = h/mc. At this stage, the universe consisted of about 99% Hydrogen and Helium atoms with the remaining 1% consisting of atoms of the 2000 or so possible isotopes of the other elements. Among these great clouds of atoms that eventually become stars and galaxies, the cosmos contained a background of 2.7°K blackbody photons combined with a separate background of high energy cosmic gamma ray photons, also with an overall temperature of approximately 3°K.

The Hubble Red Shifted Galaxies
e/p mass ratio about 1/900
Bohr radius $a_o = \alpha \lambda_\infty / 4\pi = 2.04 = 1.0796 \times 10^{-10}$ m
Fine structure constant $\alpha = 4\pi a_o / \lambda_\infty = .00175$
Hydrogen ionization photon $= \lambda_\infty = 4\pi a_o / \alpha = 7.75 \times 10^{-7}$ m
Neutron Stability Number $= M_p / M_E \sqrt{\alpha} = 37.6$

After many billions of years, most atoms in the Living Universe had gathered together into clouds, planets, stars, and galaxies. The most distant of these galaxies that we can now see with the Hubble telescope emitted red shifted photons with wavelengths of about $Z = 7$. As electron mass constantly decreased, it caused both the Bohr radius and the fine structure ratio to decrease proportionally. This increased the energy and temperature at which atoms emit blackbody spectral photons. At this stage in electron evolution, the spectral photons emitted by atoms with heavier electrons had wavelengths that were about 8 times longer than the same spectral photons we measure from atoms here on Earth. The galaxies all remain relatively stationary and are not expanding away from one another. The Hubble red shift is caused by the individual expansion of electrons and not the general expansion of the cosmos.

The Dark Energy Myth
e/p mass ratio 1/1600
Bohr radius $a_o = \alpha \lambda_\infty / 4\pi = 1.148 = 6.075 \times 10^{-11}$ m
Fine structure constant $\alpha = 4\pi a_o / \lambda_\infty = .0055$
Hydrogen ionization photon $\lambda_\infty = 4\pi a_o / \alpha = 1.39 \times 10^{-7}$ m
Neutron Stability Number $= M_p / M_E \sqrt{\alpha} = 119$

It is from this era of the Living Universe that astronomers have recently observed that the most distant measurable supernova explosions have considerably less energy and intensity than supernova explosions in nearby galaxies. Foolish Big Bang theorists tried to explain this energy loss with a whole new dimensional layer in their spacetime continuum. Unlike the Guth layer of reality that just instantly expanded and then disappeared, this new layer takes the form of an increasing antigravity force field that is somewhat stronger at long range than Newton's universal gravitation and causes the galaxies to slowly accelerate toward the outer edges of the cosmos. This imagined antigravity repulsive force has been called everything from Quintessence to Dark Energy.

The whole bone-headed idea of Dark Energy is nothing more than an ad hoc assumption by Einstein's followers to validate his idea of an eternally constant 1/1836 electron/proton mass ratio. However, these well established astronomical measurements of decreased supernova energy and intensity are not at all unsuspected in the Living Universe.

The so called Dark Energy effect is simply the expected result of decreasing electron mass. At this point in e/p ratio time, supernovas emitted spectral photons with less energy and longer wavelengths than the same photons today. These less energetic photons decreased the overall energy and intensity of supernovas from that time period.

Dinosaurs Ride Along on the Continents
e/p mass ratio 1/1800
Bohr radius $a_o = \alpha \lambda_\infty / 4\pi = 1.02 = 5.506 \times 10^{-11}$ m
Fine structure constant $\alpha = 4\pi a_o / \lambda_\infty = .00702$
Hydrogen ionization photon = $\lambda_\infty = 4\pi a_o / \alpha = 9.9 \times 10^{-8}$ m
Neutron Stability Number = $M_P / M_E \sqrt{\alpha} = 151$

At this stage of the Living Universe, the evolution of the cosmic e/p constant is able to answer some long contemplated mysteries in Earth's geological history.

The biological and physiological study of the largest of dinosaurs has long shown that they were much too large and heavy to have a viable existence on today's Earth. Modern physiology predicts that their muscles and bones could not have supported the great weight of largest dinosaurs.

The simple answer to this paradox is that when the e/p ratio was at 1/1800, large dinosaurs were able to easily walk, run, and maneuver because surface gravity was much less on a larger and less dense Earth. Earth had the same mass as today, but because it was larger, its surface gravity was less. The decreasing mass of the electron and the decreasing dimension of the Bohr radius causes atoms to slowly become smaller and Earth to become denser.

If Earth's radius was twice what it is today, its density would be one/eighth and its surface gravity would be one/fourth (2.4 m/s^2) This is almost identical to the gravity of the Moon. Certainly, dinosaurs that are too heavy to be able to walk around on Earth would be able to get up and run on the Moon.

Plate Tectonics and the Shrinking Continents

Atoms are held together at the Bohr radius. The Bohr radius link that holds the electron and proton together in the Hydrogen atom is considerably larger than the 79 Bohr radii links that hold a gold atom together.

As Hydrogen atoms get smaller and emit more energetic photons, the atoms of all the chemical elements also decrease in size and increase in spectral energy by a lesser amount. The larger an atom's mass and number of electrons, the less its size decreases in proportion to the hydrogen atom. Atoms at the heavy end of the periodic table shrink considerably less than the atoms at the light end of the table.

When Earth was in a molten state, the heavy elements tended to sink toward its center, while the lighter elements floated to the surface. Once Earth had cooled to a semi-solid state, cracks began to develop in its outer crust composed of mostly elements lighter than Iron, such as Silicon and Oxygen. As the light elements in the outer layers of Earth's mantel contracted faster than the heavy elements in underlying layers, cracks open up in Earth's continents and they appear to move apart in the observed processes of sea floor spreading and plate tectonics. This is an optical illusion. Earth's surface layers are actually shrinking and cracking like mud on a drying lake bed.

In the Living Universe, plate tectonics is a natural process of electron evolution and not a theory. The motion of Earth's continental plates is a conclusion of measurement and does not require the initial assumption of a theory.

The way this process works is that as electron mass decreases, the fine structure constant ratio grows smaller. This in makes the Bohr radius and thus the size of atoms grow smaller. It is concluded that the rate by which the physical size of a particular atom decreases is dependent on the number of its bound electrons.

This effect provides an easy answer to one of the most difficult of Earth's geological mysteries. Geologists have long tried to supply a mechanism to explain the apparent break up and spreading apart of Earth's continents. Plate tectonics is the latest idea to explain this phenomenon but it sometimes presents more questions than it can answer. Even if all the evidence for continental drift could be explained by the movement of large plates in Earth's crust, there is still no underlying mechanism that can make the plates move apart in the first place.

There is a great deal of geological evidence to support the idea that Earth once had a single large unbroken continent called Pangea. Between then and now, Pangea broke apart into a number of continents and islands that drifted apart over much of Earth's surface. The experimental evidence for this event is excellent. There have long been examples that made Earth and several other heavenly bodies appear to be expanding with surfaces cracking and continents appearing to be moving apart. The problem is, no one has been able to come up with a physical system that can come close to making the whole process work

In the Living Universe, the shrinking of the lighter atoms in Earth's crust at a faster rate than the heavy atoms in its interior causes the surface to crack apart as it contracts faster than the heavier atoms in Earth's interior. There is no mystery here. This is exactly what we would expect from an increasing e/p mass ratio

Today, the Evolution of Matter in the Living Universe Continues
a/p mass ratio 1836/1
Bohr radius $a_o = \alpha \lambda_\infty / 4\pi = 5.292 \times 10^{-11}$ m
Fine structure constant $\alpha = 4\pi a_o / \lambda_\infty = .0073$
Hydrogen ionization photon $= \lambda_\infty = 4\pi a_o / \alpha = 9.11 \times 10^{-8}$ m
Neutron Stability Number $= M_p / M_E \sqrt{\alpha} = 157$

Today, we use our instruments to measure the momentum and angular momentum in the Living Universe from our moving position on Earth. Historic measurements of the electron/proton mass ratio of 1/1836 are beginning to show, that today, the ratio has increased to 1/1836+? As the electron's energy/mass continues to decrease, the increasing momentum and ionization energy of atoms is continually being converted into photons with more energy and shorter wavelengths. This causes both campfires and stars to get hotter and warm the cosmos. However, the universe's total increase in temperature from radiating stars is still only a tiny fraction of the original 2.7°K temperature of the Cosmic Blackbody Radiation.

Today, in the Living Universe, all of the Hubble galaxies are emitting spectral photons that are identical to photons measured here on Earth. When these photons eventually reach us, we will measure them as red shifted because during their travels the cosmos has become increasingly blue shifted.

Since the time of the original anti-Hydrogen atom nearly half of the rotational energy/mass of the cosmos has been converted into the linear energy/mass of photons and moving atoms. All of the mass that was lost in the electron's evolution is still conserved in the energy/mass of photons and the kinetic energy/mass of moving atoms.

Dynamic interactions within the cosmos are constantly converting rotational energy/mass into the linear energy/mass of photons with decreasing momentum, energy, and longer and longer wavelengths. This is the second law of thermodynamics. Entropy is photon energy lost to the void. The universe's cooling effect of entropy is balanced by the warming effect of the decreasing wavelengths and increasing temperature of thermal radiation photons. It is very likely that the total entropy of the universe is neutral.

When atoms first formed, during the Grand Fire, the neutron stability number was slightly greater than one and the neutron was virtually stable. At that e/p ratio time interval, the vast majority of the element's approximately 2000 possible nuclear isotopes were at least virtually stable. Then, as electron mass gradually decreased, the steadily increasing value of the neutron stability number decreased the neutron's half-life. As a result, once stable isotopes began decaying, one after another. Today, with a neutron stability number of 157, the chemical elements only have 282 stable isotopes left. In the distant future, there will be less stable isotopes as the Living Universe continues to evolve. As neutrons become less and less stable there will come a time at the end of the world when they cease to exist and the universe will contain only 2^{256} hydrogen atoms and countless photons moving out into the void in all directions.

It will be a very long time before there are no longer enough stable isotopes for humans to get by on. I would think the metallic isotopes like Gold and Palladium would disappear long before we have to worry about losing any of the lighter isotopes vital to our bodies like carbon and oxygen. Long after we are gone, and the last alpha particle has decayed, the Living Universe will die a quiet death as a cloud of pure hydrogen gas containing exactly 2^{256} atoms. The universe began as a solitary ground state anti-hydrogen atom with a mass of one and no kinetic energy. The Living Universe will end as 2^{256} individual ground state atoms each with a mass of $1/2^{256}$ and no kinetic energy. There will still be a few photons around but most will have been lost within their one-dimensional voids. All is conserved.   *One day it will all end. Enjoy it while you can.*

The Inventions of the Big Bang Explosion
All Big Bang theories begin with the metaphysical assumption of the eternal constant of 1/1836 for the electron/proton mass ratio. From this they conclude that the universe's matter and energy exploded from some imagined point-like location called a singularity. Then, as the singularity expanded and "cooled" to a point where the temperature was just right, a number of protons appeared from the cooling residue of the singularity's photons. Then, after an even longer period of time a similar number of electrons spontaneously appeared within the expanding spacetime and the singularity was gone. From this time on, no electrons or protons could be spontaneously created from photons without their antiparticles and no protons or electrons could disappear into photons without their antiparticles.

Let there be Atoms
The first known invention of an instant creation process occurred when the author of Genesis wrote "Let there be Photons". This simple statement is still the basic assumption for Big Bang's singularity. The problem with this assumption is that it is upside down. The first law for the creation of photons is *Let there be Atoms*. Photons are secondary to atoms. Atoms emit and absorb photons. Photons do not emit and absorb atoms. The singularity is upside down and backwards because it tries to make atoms out of photons instead of the making photons from atoms. If we begin the Living Universe with atoms instead of photons, we can arrive at today's cosmos without making any other metaphysical assumptions or violating the accelerometer measurements of mass, space, time, and gravity.

The really weird rule followed by theoretical cosmologists is that they are totally free to construct any number complicated and convoluted theories to explain a process by which vast amounts of pure photon energy could have appeared from nowhere at the beginning of the universe and then gradually transformed into the atoms and photons of today. By contrast, it is strictly taboo for cosmologists to propose that only atoms appeared from the singularity and that the only photons back then were those emitted by atoms. This is an electrodynamic law that all cosmologists fail to heed.

Any experimental physicist will tell you that it is very easy to get photons from atoms but nearly impossible to get atoms of matter from photons. Certainly, electron/positron pairs and proton/antiproton pairs can be produced from photons, but it is difficult to make atoms out of them and when you do, you make equal numbers of atoms and anti-atoms. If these come in physical contact they will annihilate back into pairs of photons with half the energy of the original photon. Experimental physicists have long known that you cannot create a large quantity of stable matter from photons without physically separating the antimatter from the matter.

Astronomical measurements show that the cosmos contains almost no positrons and antiprotons. The best example of this is the measurements cosmic rays. Cosmic rays give us a very good picture of the physical contents within the cosmos at large. Cosmic rays consist of photons and high speed particles of matter coming from all directions. They are mostly electrons, protons and alpha particles but there is a sprinkling of stable nuclear isotopes from all the elements. The abundance of individual isotopes in cosmic rays is quite similar to the relative abundance of elements here on earth. If there were any quantities of positrons and antiprotons within the cosmos, they would show up in cosmic rays. Believing that the cosmos could have been created without antimatter is an impossible thing.

Other Beliefs in Impossible Things
Experimental physicists measure all possible things and theoretical physicists imagine all the impossible things that cannot be measured.

The first impossible things imagined by all theoretical physicists are the spacetime aether fields and the continuum dimensions. Theorists define one or more of these to explain non-local forces and motion but these forces are impossible to measure by experimentalists. For example, imagined gravitational fields are believed to produce downward forces but no such forces have ever been measured.

Big Bang theorists imagine a spacetime field in their theories and then go on to invent several more similar but distinctly different entities to explain the features of their imaginary and impossible cosmic creations such as Dark Energy and the cooling of the Cosmic Blackbody Radiation.

Besides the pure energy singularity itself, the next big impossible thing they come up with is the Guth inflation. Inflation was invented and developed by impatient cosmologists who did not want to wait the billions of years that it would take for the galaxies in the universe to move from their point of creation to their present locations. Instead of waiting, theorists spread matter and photons throughout the universe in a virtual instant of time with Alan Guth's idea of matter teleportation. Guth inflation is an impossible thing that violates the constant speed of light and all other physical laws of force, acceleration/deceleration, momentum, energy, and gravity.

Perhaps the favorite impossible thing believed by all Big Bang enthusiasts is the idea that 2.7°K Cosmic Blackbody Radiation has cooled from an original temperature of about 3000°K. Experiments show that their is no way that blackbody radiation can be cooled without destroying its blackbody distribution curve. The physical way to cool radiation is to move the photons farther apart. Since this changes their distribution curve, Big Bangers believe photon wavelengths are constantly increasing while their momentum and energy are decreasing by proportionate amounts. The most amazing impossible thing about this expanding space is that it has three different kinds of expansion that all happen at the same time. In the first type of expansion, the wavelengths of photons increase, and their momentum and energy decreases. In the second type of expansion the distance between photons is increased to match the blackbody temperature curve. In the third type of expanding space, the space between the galaxies expands but the space within matter and galaxies does not expand. Both photons and the universal void of space expands while atoms, stars, and galaxies remain a constant size. It seems impossible that this imagined expanding space could cause some things to expand and not others. Also, how is this expanding space able to absorb more than 99% of the momentum and energy ever produced in the cosmos and have no effect on the speed of light or the energies of atoms or stars?

There is a strong controversy among Big Bang buffs as to the true cause of the Hubble red shift. One groups says it is left over outward velocity from the original Guth inflation and the other group claims the galaxies are inertially stationary and it is the CBR expanding space that is increasing the wavelengths and decreasing the momentum and energy of the Hubble photons. The problem here is that the Hubble photons have increased in wavelength by a maximum of about 10 times and the CBR photons have increased by about 1000 times.

The cosmologists next big impossible thing is the idea of a new type of spacetime continuum called Dark Energy. Unlike the CBR expanding space, this new Dark Energy is a repulsive force that accelerates all matter away from the center of the cosmos. The only reason for these ideas and beliefs about Dark Energy and other impossible things is the cosmologists' unquestioned faith in the eternal 1/1836 electron/proton mass ratio.

It seems the standard solution for all modern cosmology problems and paradoxes is for theorists to immediately invent a new spacetime continuum field and adjust its parameters to solve each new problem. There seems to be no limit to the number of new and exotic spacetime aethers that Big Bang theorists can imagine. The only way to make progress in theoretical physics is to eliminate impossible metaphysical assumptions entirely and concentrate on the physical assumptions made by experimental physicists.

The Big Bang Theory Begins

It wasn't until well into the 20th century that George Lamaitre began to apply scientific principles to the idea that the universe had a sudden beginning that was created by God. His initial idea was essentially to create the universe from a single giant atom that somehow split apart into the atoms of today. George had the idea of the Living Universe basically figured out but since he didn't know about photons, neutrons, antimatter, Hubble red shifts, or the 2.7°K CBR, he was unable to fill in many of the details.

However, his followers soon took his ideas of creation and turned them around into an expanding universe model that began with the arbitrary idea of a Big Bang singularity that created atoms from photons. This was followed by a long list of metaphysical assumptions describing parameters and principles that painted a somewhat coherent picture of a creation of atoms, stars and galaxies. What came to be called the Standard Model Big Bang Theory contained many contradictions and violations of the physical laws of nature but these were explained away with such ideas as complementarity and relativity. The standard model of the Big Bang combines astronomical measurements with a large number of theories and contradictory assumptions about the existence of unmeasured parameters and certain imagined ancient laws of physics that are no longer in effect today. All of the Big Bang theory's structural problems stem from its assumptions of eternal electron mass and the idea that atoms can be constructed from photons.

According to some quantum mechanical Big Bang inventions, 2.7°K is regarded as a random temperature point in the cooling process of the CBR that began long ago in a much hotter and denser cosmos. In the circlon model of atomic structure, 2.7°K is the only possible temperature for this predicted homogeneous event that transformed a cosmos full of neutrons into atoms radiating photons.

Big Bang Theory's Violations of Natural Laws

There are many ways that Big Bang theorists violate the laws of experimental physics by proposing natural phenomena that cannot be measured or even detected. Of these, there are three in particular that have never been satisfactorily resolved by any experiment.

The creation of matter without antimatter. No experimental physicist has ever been able to create a particle of matter without also creating an equal particle of antimatter. Why do we not detect any leftover antimatter anywhere in the cosmos from the initial creation of our protons and electrons?

The answer is that what is called "matter and antimatter" is really positive magnetic matter (protons) and the negative electric matter (electrons). These matter and antimatter particles are still with us in the Living Universe today in exactly equal numbers. The electron was once the antiparticle to the proton when their masses were equal.

The cooling of the Cosmic Blackbody Radiation without any transfer of energy to the rest of the cosmos at large. How is it possible for the CBR to cool if momentum and energy from its photons are not transferred to the rest of the matter in the cosmos? Where does this energy go? The law for the conservation of energy does not allow heat to disappear. The CBR could not cool without heating up something else. Big Bang people claim the CBR photons originally had 1000 times more momentum and energy than they have today. Where did all of this non-conserved energy and momentum disappear to?

The answer is that the CBR had a temperature of $2.7°K$ when it was formed and it still has the same temperature today.

Resolving the paradox between the Doppler and non-Doppler shifts of the Hubble red shifted photons and the Cosmic Blackbody Radiation. When we look at photons from the cosmos, they comprise two distinct groups. More than 99% make up the CBR and the other less than 1% are the photons produced by atoms, stars, and galaxies. The first group consists of blackbody photons with a single temperature of $2.7°K$ and the second group consists of mostly spectral photons from atoms at all possible temperatures. Both the CBR photons and the highest red shift Hubble photons from the edges of the universe are assumed to have been emitted from the same atoms during the first billion years of cosmic history. The Big Bang theory assumes that all of the photons from both groups were emitted from atoms with temperatures of around $3000°K$.

One of the great paradoxes in Big Bang theory is how the Hubble group of photons acquired only modest red Doppler shifts of less than $Z=10$, while the CBR photons acquired enormous red shifts of $Z = 1000$. According to theory, both groups traveled a similar distance through the same space during nearly 14 billion years since the singularity was predicted to have occurred.

The solution to this paradox is that neither group of photons is Doppler shifted and both received their "shifts" from the evolving energy/mass of the electron. The CBR photons were emitted from a time much further in the past than were the Hubble photons.

The Electrodynamics of Absolute Circlon Synchronicity

The main feature of circlon electrodynamics calculations is that Planck's constant $h = m\lambda c$ is not a single metaphysical constant but a combination of the two physical constants of photon masslength $Y = m\lambda$ and the speed of light $h = Yc = m\lambda c$. This eliminates the need for the massless photon as well as the transformation between mass and energy as in $e = mc^2$. The correct form of this photon equation is $cC = e/m$. The photon's equal linear and rotational kinetic energies are $e = mc^2/2 + mC^2/2 = mcC$. Mass is the absolute and constant measured component of energy $e = mv^2/2$, momentum $p = mv$ and angular momentum $I\omega = mvr$.

The energy of the photon $e = hf = m\lambda cC/\lambda = mcC = mc^2/2 + mC^2/2 = E$

The angular momentum of the photon $I\omega = h/2\pi = m\lambda C/2\pi$

Today, when the Bohr radius is $a_o = 5.2 \times 10^{-11}$m and the fine structure constant is $\alpha = .007$, the electrodynamics of the Hydrogen atom produces its intrinsic Lyman spectral photon $_{Ly}\lambda_\infty = 4\pi a_o/\alpha$ at a wavelength of 9.11×10^{-8}m. This is the shortest possible wavelength in the hydrogen spectrum and when it is emitted, it leaves the Hydrogen atom at its ground state.

The angular momentum of the ground state Hydrogen atom is. $I\omega = m_e a_o \alpha C = 1.06 \times 10^{-34}$ This constant for angular momentum is the same for all photons $I\omega = m\lambda C/2\pi = 1.06 \times 10^{-34}$. This value is a universal constant because it is not changed by the evolution of electron mass. As electron mass decreases, the fine structure constant α increases and the Bohr radius a_o decreases to maintain a constant value for this so called "quantum" of angular momentum. This is the angular momentum at an atom's Bohr radius as well as the angular momentum of all photons. An atom, must have at least this quantity of angular momentum $I\omega = h/2\pi = M_e a_o \alpha C$ between its proton and electron in order to emit a photon. The wavelengths of spectral photons are transformed as decreasing electron mass causes changes in the fine structure of the Bohr radius. These changes are required by the conservation of angular momentum.

These electrodynamics explain the value of the Hubble shift and are also able to calculate both the 2.7°K temperature as well as the precise timing of the of the initial burst of 2.7°K cosmic blackbody photons. As electron mass decreases, it increases the fine structure of a decreasing the Bohr radius. This, in turn, decreases the wavelengths of spectral photons. Decreases in an electron's mass within the structure of a neutron decreases the neutron's stability and increases its decay energy.

Physics of the Big Bang without the Metaphysical Assumptions
The Living Universe principle of cosmic creation is not a theory. It is a series of conclusions made from measurements that are fitted together into a theory-like structure for the evolution of the electron within the living and growing cosmos.

This is not meant to be a new theory of matter, energy, or even cosmology because none of the initial metaphysical assumptions of a theory are made prior to the conclusions of experimental measurements. It is just a new way of looking at Big Bang physics that is upside down, inside out, and backwards from the standard model theories based on massless photons, downward pointing gravity, and the eternal and constant value of the 1/1836 electron/proton mass ratio.

The value of 1/1836 can be used to represent an interval of cosmic time on an Electron Evolution clock. $T_{EE} = 1/1836$ is the time interval that we are living through today. The next interval of Electron Evolution Time $T_{EE} = 1/1837$ will be in our future when the electron/proton mass ratio has increased to that amount. Both campfires and the sun will have become slightly hotter as well as the neutron becoming slightly less stable.

In our journey back into the far reaches of a Living Universe, we begin at this future interval of time and then stop briefly in the present at $T_{EE} = 1/1836$. We then continue our journey into the past with stops at $T_{EE} = 1/1800$, $T_{EE} = 1/900$, $T_{EE} = 1/147$, $T_{EE} = 1/146.5$, and finally at $T_{EE} = 1-/1$. From there, we go backwards in Electron Evolution time from $T_{EE} = 1+/1$ to $T_{EE} = 147/1$, and finally to the beginning of the Living Universe at $T_{EE} = 1836/1$.

A Beginning of Cosmic Time at $T_{EE} = 1836/1$
At this arbitrary point of beginning, today's negative electron and positive proton have been transformed by the reverse evolution of matter into a negative antiproton and positive positron with an antiproton/positron mass ratio of a/p = 1836/1. This point of beginning was when the positron and antiproton coupled together into an Anti-Hydrogen atom and emitted the photons necessary to reach its ground state.

$T_{EE} = 1836/1$ is not meant to be a beginning of cosmological time. It is simply the most logical and symmetrical point in time to start a complete evolutionary story of a Living Universe from beginning to end. It begins a detailed description of the existence and evolution of the atoms of matter and photons. In this initial condition of the anti-cosmos, only one anti-Hydrogen atom existed with an antiproton/positron ratio of a/p = 1836/1. As we now go forward in time, the anti-atom sits at rest and is otherwise dormant except for the decreasing mass of antiproton evolution. When the antiproton/positron mass ratio reaches T_{EE} = 146/1, the tertiary coil of the antineutron's circlon shape becomes larger than the secondary coils of the positron's circlon shape. When this happens, the positron spontaneously collapses and becomes locked inside the structure of the antiproton. This process is called "positron capture" and forms a stable antineutron.

The evolution of the Living Universe begins with a single assumption that can be called a metaphysical assumption because it is open ended to time. This initial assumption is that the positive matter protons and the negative matter electrons are eternal and have always existed in equal numbers within the void of space. Their motions and interactions are measured to be in compliance with the standard Newtonian laws of physics. A conclusion of measurement made from this assumption is that the electrons (negative matter) have been slowly evolving in a synchronous process that decreases their energy/mass and increases their size (wavelength).

It is concluded that a Living Universe could begin with a single anti-Hydrogen atom that had existed long before this arbitrary "beginning". The only other assumption that need be made is that this original anti-atom contained all of the energy/mass in today's universe. We do not need to explain or even question the existence of the original positron and antiproton at the beginning. They were made of the same quantity of negative electric matter and positive magnetic matter that still exists in the Living Universe today. We do not need to assume the existence of electric and magnetic matter back at this arbitrary beginning any more than we need to assume their measured existence today. Once we conclude that matter has always existed, we can explain the history and structure of our world without any new laws, assumptions or parameters being added to the already well established laws of experimental physics.

Points in Space and Time

Philosophically, the idea of a beginning of time is kind of an oxymoron. Time is experienced through the thought process and while thoughts have beginnings and ends, a beginning or end to time cannot be imagined. The same is true for the ideas of the point and infinity. We cannot logically begin the cosmos at either a point in space or a point in time. Points in space and a beginning of time are even more difficult to imagine than infinity. At least you can look far into the heavens with a telescope and imagine you can see infinity. However, a point cannot be seen or even imagined with even the most powerful of microscopes. From this, we can only conclude that the universe could not begin as photons and particles appearing with a "bang" from a point within a spacetime aether that had always existed. The Living Universe always contained individual circlon shaped particles of electric (negative) and magnetic (positive) matter within a featureless void.

To discover the true nature of what was happening to electrons and protons during their reproduction, we must begin by studying what is happening in the interaction of matter today and then work our way back into the past toward this event. In this way, we can examine the forensic measurements of matter's evolution in terms of physical laws rather than make metaphysical assumptions about imagined and unmeasured initial conditions for the beginning. The only evidence presented here is the measurable dynamics of matter and photons. From these first principles of measurement, we can trace the evolution of matter back to its existence long before the so called Big Bang singularity.

The first clue in our quest for the evolution of matter is the discovery that the mass of the electron is slowly decreasing while its size has been increasing by a proportionate rate. This decrease has been detected in a general way by measurements of electron and proton mass going back to their discoveries at the turn of the 20th Century. This is an experimental measurement that we can make here on Earth. I have no doubt that if we develop the technology to make extremely precise measurements of proton mass, electron mass, the Bohr radius, and the fine structure "constant", it will not be too many years before we will be able to detect and then measure the rate of electron evolution in the laboratory.

When we look away from Earth and point our telescopes deep into the cosmos, the Hubble red shift becomes the first independent confirmation of this discovery. It shows us that spectral photons emitted by atoms today have much shorter wavelengths than the same spectral photons emitted in the distant past.

We must not make any assumptions or theories about the cause of the Hubble shift and instead accept these measurements at face value. What these shifted photons obviously tell us is that atoms in long ago galaxies emitted spectral photons with less momentum and longer wavelength than they do today. This is what we measure, but not what Big Bang cosmologists want to believe.

These theorists believe these large red shifts are Doppler shifts caused by distant galaxies rushing into the void at speeds approaching the speed of light. However, to propose such an idea without any collaborating evidence seems quite preposterous. When cosmologists hear hoof beats in the distance, they will immediately insist that it has to be metaphysical unicorns. The theorist has the choice between trying to account for the tremendous energies of an exploding universe or just calculating the changes in energy produced by evolving electron mass. The experimental physicist must make conclusions from what can be measured while the theoretical physicist is free to imagine impossible things that cannot be detected.

Unless one makes the initial assumption of eternal electron mass, there is no logical reason to conclude that the Hubble shifts are Doppler shifts caused by the rapid motion of distant galaxies. They are simply the electrodynamic effects of expanding electrons and not the Doppler shifts of an expanding universe.

Measurements conclude that in the past, atoms emitted photons with longer wavelengths than they do today. The Hubble shift does not require any special explanation because such a shift is required by electron energy/mass transformation. The reason for the cosmological red shift is that as the electron's energy/mass decreases, the electrodynamics of atoms require them to radiate photons with shorter and shorter wavelengths.

The Hubble shift and the circlon shapes of electrons, protons, and neutrons are all we need to trace the evolution of matter and energy back to its earliest beginnings without inventing metaphysical assumptions or theories that are not supported by today's measurements of quantum mechanics or electrodynamics. Matter's cosmic evolution is driven by decreasing electron mass m_E that in turn decreases the Bohr radius a_o with an accompanying increase in the fine structure constant α. These three values change in a complementary way in order to maintain the universal value of the atomic angular momentum constant ($I\omega = h/2\pi = M_E a_o \alpha C$).

The Beginning of the Living Universe $T_{EE} = 1836/1$

The Living Universe began with two fully formed conscious deities. The antiproton and positron had existed forever separately and have just joined together to form a single atom of anti-Hydrogen. We could go further back in time but this single anti-atom can tell us everything that we need to know about the evolution of the Living Universe.

The antiproton/positron mass ratio was 1836/1 and they were identical in structure to the anti-Hydrogen atoms made in the laboratory today except that they contained the entire mass/energy of the universe. (Mass = 10^{80} protons, $M_p = 1.76 \times 10^{-27}$ kg = 10^{53} kg total. Energy of the universe e = mc^2 = $10^{53} \times 10^{17}$ = 10^{70} Joules). All of this energy was rotational energy/mass. At this time, the cosmos contained no linear energy/mass and as a result its exact temperature was absolute zero 0°K.

The positron was coupled to the antiproton as they quietly sat alone at the anti-atom's ground state. They occupied the zero momentum position of absolute rest within the void of space that we can easily imagine to be its center. All atoms are at rest within this frame and have no momentum and all photons move at the speed of light c relative to this zero momentum rest frame and no photons are Doppler shifted in this frame. The only change that this atom experienced with the passage of time was the negative electric matter of the antiproton slowly decreasing in energy/mass and increasing in size. Experimental physicists have just barely detected the increase e/p mass ratio since it was first measured to be 1/1836 at the beginning of the Twentieth Century. Today, the ratio is measured to be about 1/1836+?.

Positron Capture $T_{EE} = 146/1$

As long periods of time passed, the rotational energy/mass of the antiproton decreased while the energy/mass of the positron stayed constant. In the process,
the lost rotational energy/mass of the antiproton becomes the linear energy/mass of vibration between the two particles. Eventually, the antiproton/positron mass ratio dropped to 146/1 and the tertiary coil of the positron's circlon shape decreased to $a_o\sqrt{\alpha} = 1$. This caused positron's primary coils to be pulled inside of the antiproton's secondary coils. This process of positron capture transformed the anti-atom into an antineutron.

Bifurcation Synchronicity

As the linear energy/mass between the two particles within the antineutron increased, a breaking point was finally reached and it bifurcated into two identical antineutrons that moved apart with the stored up linear energy/mass of the original particle. The two new antineutrons moved away with relative linear energy/mass but had no internal linear energy/mass. As the rotational energy/mass of antiproton was slowly transformed into the linear energy/mass of the two antineutrons it again reached a breaking point and both particles bifurcated simultaneously into four antineutrons, all at 0°K with no internal linear energy/mass.

The four neutrons duplicated in the second bifurcation had shorter lifetimes than their parents because the circlon shapes of the antiprotons and positrons were closer to synchronicity in mass and size.

The first few neutron bifurcations cycles took billions of years, but they occurred with perfect synchronicity so that even though they were billions of light years apart and rapidly moving in all directions they all bifurcated simultaneously. Each pair of new antineutrons is exactly like its parent with 1/2 the energy/mass and twice its size. As this process continued the number of particles in the cosmos was increasing by a power of two 2, 2^2, 2^3, 2^4, 2^5, 2^6, 2^7, etc.

As the circlon shaped particles grew closer and closer to mass/size synchronicity, the antineutrons gradually decreased the time intervals between bifurcations, making the lifetimes of the antineutrons shorter and shorter. By the time of the 256th bifurcation cycle, there were less than seconds between cycles and the particles were ejected from each other with very little linear energy/mass

At this point in cosmic time $T_{EE} = 1/1$, the reproduction process stopped when the antiproton/positron mass ratio went from 1+/1 to 1/1+ and it conceptually became the electron/proton mass ratio.

The Frozen Cosmos e/p = 1/1+ thru 1/146

At the point in time when the bifurcations stopped, the cosmos consisted of 2^{256} stable neutrons. I chose this number because it has great symmetry and is just slightly less than Eddington's venerable estimate of 10^{80} protons for the mass of the cosmos.

These particles can be considered as frozen because they had very little relative motion between them and there was very little vibrational motion within the particles.

It can be pointed out that at the moment $T_{EE} = 1/1$ when the positrons were conceptually transformed into protons, the cosmos contained exactly equal numbers of matter and antimatter particles, and all particles had equal quantities of energy/mass. The negative electron (antimatter) is the primordial antiparticle to the positive proton (matter). While the numbers of negative and positive particles have always been equal, this was the only moment in time when particles of matter and antimatter contained equal quantities of energy/mass.

In Einstein's Big Bang spacetime creation theory, neutrons and antineutrons play little or no part in his imaginary ideas about the creation of matter. Theorists believe that protons were created just after the singularity appeared and electrons were not created until much later in the process. There is nothing in their reasoning requiring the cosmos to have equal numbers of electrons and protons even though the law of charge conjugation does demands equal numbers.

In experimental nuclear physics, the only process that can produce equal numbers of protons and electrons is the decay of neutrons. The Big Bang's creation model of matter without antimatter has no experimental verifications. No Big Bang enthusiast has ever offered a credible reason as to why their Big Bang universe contains no antimatter or why electrons and protons seem to be in equal numbers..

During the millions of years that the 2^{256} stable neutrons floated and moved through space, the rotational energy/mass of the electron continued to convert to the internal linear energy/mass of the neutrons. Unlike the antineutrons, this increasing energy/mass did not cause the neutrons to decay because the mass/size ratios between the particles was growing larger instead of smaller as with the antineutrons.

When the electron/proton mass ratio finally reached 1/146.5, the primary coils of the electron's circlon shape had grown slightly larger than the proton's secondary coils. Prior to this, the electron and proton were coupled together in a stable neutron with the electron's internal structure locked firmly inside of the proton. As soon as the electron's primary coils of became too large to fit inside of the proton's secondary coils, they popped out of the proton with the tremendous amount of linear energy/mass that had been stored in the particles from their long incubations.

After $T_{EE} = 1/146.5$, the electron could attach to the outside of the proton at the Bohr radius to form an atom, but it could no longer attach to the inside of the proton to form a neutron without the addition of external coupling energy. After this point in time, neutrons all become unstable and decayed into electron and protons. These electrons and protons could then either couple together into atoms and emit photons or with enough collision energy they could transform back into slightly unstable neutrons. Since then, free neutrons have gone from being virtually stable to today's lifetime of about 19 minutes.

The 2.7°K Frozen Fire $T_{EE} = 1/147$

After their long incubation process, the neutrons had accumulated tremendous amounts of kinetic energy from the decreasing energy/mass of the electron. When the neutrons finally decayed all at once, they were like tiny atomic bombs. A long frozen cosmos suddenly erupted into a gigantic fire engulfing the whole universe. 2^{256} at rest neutrons were transformed into 2^{257} high velocity electrons and protons moving in all directions. These immediately began colliding with one another, emitting photons, forming atoms and converting back into nearly stable neutrons. In a mere instant in cosmological time, the universe was converted from great and small clouds of basically frozen neutrons into a seething mass of highly energetic electrons, protons, and neutrons.

Nuclear Synthesis $T_{EE} = 1/148$

In this high-energy and neutron-rich environment, protons, electrons, and neutrons collided and joined together into atomic nuclei, atoms of all the chemical elements and many different chemical molecules. By the time this great nuclear synthesis had ended, the universe was composed of about 90% Hydrogen and 9% Helium. The remaining 1% of matter in the cosmos contained quantities of at least a few atoms of all the 2000 or so possible nuclear isotopes. The process finally stopped when virtually all of the universe's supply of protons, electrons and neutrons had been absorbed into the structures of about 2000 different chemical atoms.

When they were first formed, all of the isotopes were virtually stable with a nuclear stability number of one $\#_{NS} = M_P/M_E \sqrt{\alpha} = 1$. Then, as electron evolution progressed and the neutron stability number $\#_{NS}$ increased to say 10 or 20, the neutron began to become more unstable. This caused the largest and least stable of the atomic nuclei to decay into lighter nuclei.

In the time that the nuclear stability number ($\#_{NS} = 1+$) advanced to today's value of ($\#_{NS} = 156$), the 2000 original isotopes had decayed into the 285 virtually stable isotopes that make up the periodic table. Currently, elements heavier than Hydrogen and Helium make up only about one percent of atomic nuclei. Today, most of a particular element's many nuclear isotopes are unstable, but at the time $T_{EE} = 1/147$ just after the great neutron decay, they were all at least virtually stable.

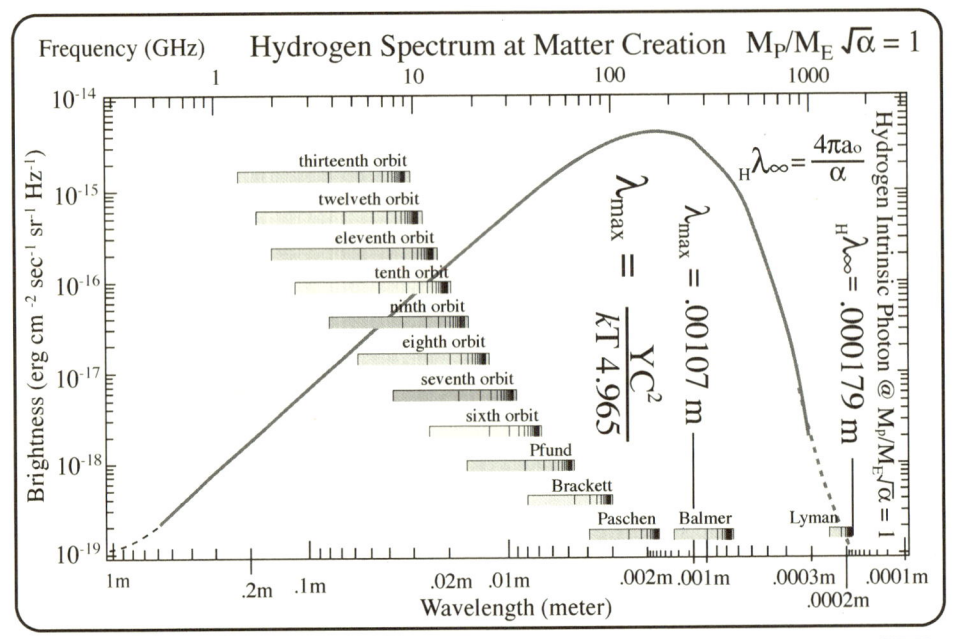

Hydrogen Spectrum at Matter Creation $M_P/M_E \sqrt{\alpha} = 1$

$_H\lambda_\infty = \dfrac{4\pi a_0}{\alpha}$

$\lambda_{max} = \dfrac{YC^2}{kT\,4.965}$

$\lambda_{max} = .00107\ m$

$_H\lambda_\infty = .000179\ m$

This drawing shows the Hydrogen spectrum compared to the CBR.

2.7°K Cosmic Blackbody Radiation $T_{EE} = 1/147$ to $T_{EE} = 1/1836$

As the process of rapid neutron decay continued, electrons began coupling to protons and other nuclei to form atoms and emit spectral photons. Both the kinetic energy of the electrons' and protons' rapid motion and the ionization energy between them was slowly converted into a mixture of spectral photons from all the elements. After this energy was dissipated into photons, the universe became filled with blackbody photons and the atoms settled down into their ground states and became relatively dormant. In this state, they possessed less than one unit of photon angular momentum and could no longer emit any photons unless they absorbed a photon or gained angular momentum from colliding with other atoms. With an electron/proton mass ratio of 1/147, heated atoms produced spectral photons at a blackbody temperature of 2.7°K.

This grand fire of radiating atoms filled the universe with a great burst of photons that contained the ionization energy of all the elements as well as much of the enormous decay energy of the neutrons. This great release of photon energy/mass can still be viewed in the universe today as the photons that make up the 2.7°K Cosmic Blackbody Radiation.

These photons represent an almost unbelievable amount of energy. Even today the CBR photons have about a thousand times as much energy as the photons produced by all the stars and galaxies since their formation.

Once the individual atoms had converted their kinetic and ionization energies into spectral photons, the universal grand fire eventually went out leaving the whole universe at a constant blackbody photon temperature of 2.7°K. The atoms continued to absorb and emit some these photons but that did not change the overall character and temperature of their blackbody radiation spectrum.

Formation of Galaxies T_{EE} = 1/147 to 1/918

The extinguishing of the grand fire began another very long and cold, era in the history of the Living Universe when the gravitational expansion of mass, space, and time began to gather the atoms into individual clouds of gas and dust with greater and greater densities. These clouds slowly became segmented into still smaller and denser clouds that eventually collapsed into planets, stars, and galaxies. As new stars formed, starlight began to light up the universe with spectral photons for the second time. This process star formation within the galaxies took many billions of years but may have been mostly completed by the time the electron/proton mass ratio reached 1/918 or about half what it is today.

Hubble Constant T_{EE} = 1/918

The galaxies with the greatest red shifts observable with the Hubble telescope at the visible edges of the universe are about $Z = 7$. From this, we can conclude that the initial formation of stars and galaxies was largely completed by the time the electron/proton mass ratio reached about 1/918. From this time in the star formation process to the present, the energies of spectral photons have increased by a factor of 8.

From this point on, the Living Universe is just what we observe it to be with our telescopes. The Cosmic Blackbody Radiation still maintains its original temperature of 2.7°K and the total temperature of the universe produced by starlight is slowly increasing but still far below 1°K. The temperature from photons originating inside the Milky Way is about 3°K but out between the galaxies it is still much colder.

Map of Cosmic Time and the Evolution of Matter

$h/2\pi = M_E C\alpha a_o$

Electron Mass M_E 1.14 x 10⁻²⁹kg	Bohr Radius a_o 5.292 x10⁻¹¹m	Circlon Constant $\sqrt{\alpha}$ .0855	Neutron Structure Constant $Q = M_P/M_E\sqrt{\alpha}$ 156.8	Fine Structure Constant α .0073	Hydrogen's Intrinsic Wavelength $\lambda_i = \frac{4\pi a_o}{\alpha}$	Hubble Z Number Z = 0
Today's Electron Mass					$\lambda = .0000000911m$	
1/1836	1.00	.0855	157	.0073	1	Z = 0
1/1800	1.02	.0838	151	.007	1.06	Z = .06
1/1500	1.22	.0701	105	.0049	1.82	Z = .82
1/1200	1.53	.056	67	.00312	3.58	Z = 2.85
1/918	2.0	.0428	39	.001825	8	Z = 7
E_M = 1/2 proton mass						
1/146.5	12.53	.00682	1.0	.0000465	1967	Z = 1966
Neutrons decay, atoms form and the 2.7°K Cosmic Blackbody Radiation fills the universe					$\lambda = .000179$ m	
1/100	18.36	.000466	.4	.000022		
Only stable neutrons No atoms or photons						
1/1						
Electron and Proton are particle/antiparticle pair						

Reverse Entropy

What we measure with our telescopes is the opposite of what cosmologists have long believed to be true. True believers in the Eternal Electron Mass Constant and the invention of the Big Bang singularity have long believed that entropy was negative and that the universe was constantly in the process of cooling down toward an ultimate "heat death". However, with the principle of electron evolution, both Dark Energy and the Hubble shift would show that total entropy is at least neutral and may even be positive. It is perhaps true that the increasing energy of spectral photons is generally equal to the energy lost from the emission of photons with longer and longer wavelengths. The Living Universe is cooling down from photon dissipation at the same time it is heating up from electron evolution.

Conservation of Momentum T_{EE} = 1/1

One of the great unsolved paradoxes of Big Bang theory concerns the simple conservation of momentum. If all of the universe's protons and electrons were ejected from a singularity point at extremely high velocities, their momentum vectors would all be one dimensional voids pointing away from the singularity point. Since adjacent particles would be traveling on virtually parallel momentum vectors, they would have almost no relative momentum or virtual angular momentum to form atoms and emit photons. They would constantly be moving farther apart from one another and there would be little chance of any of them colliding or interacting on their long journeys toward the edge of the universe. The physical mechanics of the Big Bang theory completely eliminates any physical mechanism that could gather together these speeding particles of matter into clouds, stars, galaxies, and eventually people. In Big Bang theory, all of its momentum and energy are used to send particles of matter speeding away in all directions from a single point.

In the Living-Universe, the great explosive decay of neutrons that created the electrons and protons occurred not at a single point, but from 2^{256} separate locations throughout the universe.

When they decayed, these neutrons were moving in many different directions. The energy of these decaying neutrons put all of the electrons and protons on different high energy trajectories where they could crash into one another rather than all of them moving away from the point of a universal singularity.

The Era of Zero Entropy $T_{EE} = 1/1$ to $T_{EE} = 1/1836$

Big Bang cosmologists have long adopted the eternal electron mass constant of 1/1836 as the fundamental metaphysical assumption of nature and have concluded from it that the universe is in a constant process of cooling down from a much hotter initial state. Mainstream cosmological theories have always been built around verifying this assumption of constant electron mass. However, when we look for actual physical evidence to support the conclusion of a cooling universe, we can find very little if any. All experimental evidence would indicate that the Living Universe was very cold when its atoms were formed and it has been slowly heating up ever since.

With the electron's mass getting smaller and smaller, it is easy to see why the universe is getting hotter. The photon emission spectra for all the elements are gradually growing to greater energies and shorter wavelengths. It could be that this effect matches the "heat death" predicted for the universe by the second law of thermodynamics. If the assumption is made that these effects are equal, then it can be concluded that entropy in the Living Universe is zero. As the Living Universe cools from the dissipation of energy into longer and longer wavelength photons, it is also being warmed by the increasing energy and shortening wavelengths of atomic emissions.

The Chemical Era

As the mass/size ratio between the proton and electron grows, there are subtle changes in the chemical reactions that occur from protons and electrons coupling together. Because of a much larger Bohr Radius, chemical compounds in the distant past had somewhat different properties than they do today. Perhaps the chemicals that make up dinosaur bone were stronger then than they are now.

It is possible there was a point in cosmological chemistry about three billion years ago when the chemistry was just right for the spontaneous formation of DNA and other organic molecules. Since then, these molecules have reproduced, diversified, and joined together to preserve themselves. Perhaps, there was only a small window of time in electron evolution when the chemistry was just right for the spontaneous formation of DNA molecules. After this "sweet spot" in cosmic history, DNA molecules were able to reproduce but they could no longer be produced spontaneously. All life on Earth is directly connected to these original DNA molecules. As Timothy Leary said in a lecture I once attended, "We are all the result of an unbroken chain of life that is over three billion years old." There is no place along this chain where at least some portion of our present bodies was not part of a living organism. At least in principle, our bodies could still contain atoms from one of the original DNA molecules.

In the Living Universe, the evolution of life is driven by the changing parameters of cosmological chemistry. In the future, molecules that do not exist today may be possible. Certainly, the energy and intensity of chemical reactions will change as the mass of the electron evolves.

There is another factor that may very well have a significant effect on the evolution of life in both the past and in the future. This is the increasing instability of the neutron and its link to the decreasing stability of atomic nuclei. At one time in the distant past, common and widespread radioactive nuclei such as Thorium and Uranium may have been stable and this may also have been true of many other radioactive elements and isotopes. The end result of this evolving process is that today we have some 282 stable isotopes and a small number of radioactive isotopes with very long lifetimes. As the stability of the neutron decreases with electron evolution, the least stable of these 282 isotopes will become radioactive.

Knowledge of this effect will also change the way we measure things like the age of Earth. One way to determine the age of Earth is to study and quantify the rate that Thorium and Uranium decay through a series of steps into Lead. It is possible that these weakly radioactive elements may have been much more stable when Earth was first formed. Also, there may have been many more than today's 282 stable isotopes. If Thorium and Uranium were once stable or had much longer half lives, then Earth could be much older than previously estimated based on the decay of these elements.

The Era of Conscious Thought $T_{EE} = 1/1700$ to $T_{EE} = 1/1836$

Perhaps the most remarkable thing about the whole scenario of a Living Universe is that we are here to try and figure it out and discuss it. The great unexplained mystery of the universe is not so much the existence of the chemical elements of matter but the existence and origin of consciousness.

Since the only identifiable activity in the observable universe is the interaction between electrons and protons, it must be concluded that the activity of consciousness must also be a part of these interactions. The basic units of consciousness are somehow contained within the individual structures of protons and electrons. If each of our atoms contains a basic unit of awareness, then our higher degree of consciousness results from the connections of the many atoms and molecules within our bodies. These individual atomic senses of awareness are all organized into symmetrical patterns of interactions within our atoms that we sense as feeling and experience as both consciousness and unconsciousness thought. Our total consciousness is the result of the 10^{29} possible connected interactions between the atoms in our bodies and brains.

While the atoms of a rock may have primitive individual awareness, it cannot really be called consciousness. It is possible that true consciousness did not arrive in the Living Universe until the spontaneous formation of DNA molecules about three billion years ago.

The eternal existence of matter is the initial principle of the Living Universe. A religious person, who wants to begin the cosmos with God, might want to consider concluding that electrons and protons are an eternally living Yin/Yang dichotomy of deities that could be referred to as goddesses and gods. This duality of gods has always existed with the antiproton as a yin goddess and the positron as a yang god. This god and goddess then worked together, interacted, and reproduced into 2^{256} electron and proton deities that contained all the consciousness of today's Living Universe. The god consciousness that we experience stems from the basic connected awareness of these chemical deities within our bodies.

Entropy is the evolution of total cosmic energy being irreversibly transformed from the rotational energy of atoms to the linear energy of photons and other moving bodies. In the beginning, there was no linear energy/mass and all rotational energy/mass was contained in the opposite internal motions of the original antiproton/positron pair. In the bifurcation process that divided the single anti-hydrogen atom into 2^{256} neutrons, the rotational energy/mass lost by the antiproton-electron becomes the linear energy/mass of all moving bodies in the cosmos. Entropy is a function of the general conservation of momentum, angular momentum, and energy.

Conclusion

The evolution of electron mass is not an assumption or theory because it is an experimental conclusion made from physical measurements of the cosmos without any initial metaphysical assumptions such as eternally permanent electron mass, unmeasurable parameters like a singularity, Guth's inflating field, expanding CBR photon spacetime, unmeasured Doppler red shifts, or the miraculous creation of particles of matter without particles of antimatter. Moreover, unlike the Big Bang, the Living Universe does not require the creation or destruction of any energy/mass and there is no requirement for an expanding spacetime continuum or a beginning of time. Indeed, no assumptions are made that are not completely verified by the conclusions of the scientific method for the measurement of mass, space, time, and gravity and the subsequent calculations of gravitational motion, momentum, force, and energy.

The book <u>Physics Without Metaphysics</u> contains a more complete description of the Living Universe's cosmic creation process. Each stage in the evolution of the electron and proton is laid out in terms of quantum mechanics and electrodynamics. The Living Universe model is based on physical principles and contains no metaphysical assumptions.

To purchase this, or other books by James Carter go to:

WWW.LIVING-UNIVERSE.COM

The site contains more information on Circlon Synchronicity, gravitational expansion and the circlon model of atomic and nuclear structure.

How the Pound-Rebka Experiment Shows Einstein's Two Relativity Theories to be Incompatible

When Einstein first calculated the transverse gravitational red shift of Earth's gravitational potential field, the effect was so small (2.5 x 10^{-15}) that many thought it would be impossible to measure.

It was not until 1959, four years a after Einstein's death, that Robert Pound and Glen Rebka made very accurate measurements of both direct Doppler and transverse shifts in the momentum of emitted and absorbed photons. At the time, theorists realized they had discovered and measured Einstein's long predicted gravitational red and blue shifts caused by Lorentz transformations in the intervals of clocks. Einstein believed gravitational transformations were produced by unmeasured equivalent motion whereas the Lorentz transformations in special relativity are a measure of absolute inertial motion of mass. Einstein was a fool to believe in equivalent motion because transformations in mass and time are caused by measured and not imagined gravitational inertial motion.

Pound-Rebka measures transverse and direct Doppler shifts in the absolute momentum of photons and the absolute momentum of gravity. Transverse photon shifts result from Lorentz transformations in the intervals of emitting and absorbing clocks. Direct Doppler shifts are produced by relative motion such as the movement of actuators.

Careful examination of the experimental results shows the measured shifts to be simple Newtonian transverse and direct Doppler shifts caused by changes in momentum and not by the effect of Einstein's imagined gravitational potential field or his equally implausible electromagnetic field. Photons are shifted by mass, space, time and gravity. Photon shifts can be calculated in Einstein's imaginary spacetime continuum to be exactly as measured. Einstein's calculations are correct but they are based on the metaphysical assumptions of two otherwise unmeasurable quantities.

Although Einstein's followers were able to correctly calculate the measured values of the Pound-Rebka shifts, they only did so by ignoring the Newtonian laws of force and motion.

While Einstein accepted the quantum mechanical and electrodynamic measurements of atoms and photon motion to be true, he declared the mechanical measurements of gravitational motion and force to be not equal to but only just equivalent to the force and motion of atoms and photons. Einstein's equivalent anti-principle of force and motion says that gravitational force and motion can never be measured. They are not equal to but only just equivalent to the motion and force of atoms and photons that can always be measured. The Pound-Rebka experiment clearly shows that the force and motion of gravity is equal to the force and motion of photons and not just their equivalent.

Pound-Rebka proves that gravity and inertia are equal and not equivalent.

General Relativity theorists have developed two completely different and contradictory explanations of the Pound-Rebka shifts that both calculate the same results but neither of which is likely to be true. Even though only one of these descriptions can be correct, the relativity enthusiasts declare them to be "equivalent" and thus feel justified to use them interchangeably. The actual mechanical motion and force of gravity measured by the experiment does not require either of the relativity theorists' metaphysical assumptions to explain the shifts.

The Pound-Rebka Experiment
The Pound-Rebka experiment is a pure quantum mechanical demonstration of the gravitational expansion of mass, space and time. It is a precise set of measurements of the force and motion of gravity and their interactions with the momentum of photons.

This experiment is a simple mechanical process for the measurement of transverse and direct photon Doppler shifts and does not suggest any theories of aethers, fields, or Einstein's continuum. The results can be explained completely without the need for any metaphysical assumptions of multiple dimensions or equivalent continuum fields. The interactions between gravity and photons are mechanical physical measurements of mass, space, and time. The red and blue shifts measured by Pound-Rebka are both direct and transverse Doppler shifts caused by differences in the emitter's and receiver's velocity.

In the process by which an atom emits and absorbs a photon, the photon undergoes four separate Doppler shifts depending on the quantity and direction of the atoms's absolute motion. First, it has a direct Doppler shift based on the atom's vector of absolute motion relative to the universal position of photon rest. This direct shift is different for each direction. Second, the photon has a transverse Doppler shift based on the absolute quantity of the atom's motion relative to the speed of light. Transverse shifts are the same in all directions. A third shift occurs from the observers motion relative to photon rest. The fourth shift is a transverse resulting from the observer's absolute momentum. In most cases, these four separate shifts cannot be separated from within the single measured shift. Pound-Rebka is one example where transverse and direct Doppler shifts can be experimentally separated.

Transverse shifts are absolute and have the same value in all directions. They are always a red shifts for emitted photons and equal blue shifts for absorbed photons. The transverse shift changes with each momentum increase or decrease in a body's absolute motion. Only atoms at absolute rest emit and absorb photons with no intrinsic Doppler shifts and even these photons can have a slight shift from the atom's recoil.

Pound-Rebka Physical Values

This drawing depicts the actual values for photon momentum and wavelength measured by the Pound-Rebka experiment.

Observers at the bottom and on the top of the tower observe the photons from green light bulbs located at the tower's top and bottom. The bottom observer measures the bottom green light photons as green and the top observer sees these green photons as red but sees the photons from the top green light as green even though they are measured as blue by the bottom observer.

These are the indisputable measurements of the experiment and everyone believes them to be true. Where the disputes arise is in the many ideas that theorists have to explain the physical mechanisms underlying these measured shifts in photon momentum.

Several mainstream Harvard type theoretical physicists have strikingly different explanations of these red and blue shifts and numerous crackpot theorists also have several different metaphysical theories of gravity and photon momentum.

The absolute motion explanation of the experiment presented in this book is not a "theory" of gravity or photons. It just presents physical measurements of mass, space, time, and gravity that require no theories.

Metaphysical theories are only needed when theorists like Einstein and other crackpots want to change the interpretation of the measurements in ways that cannot be detected

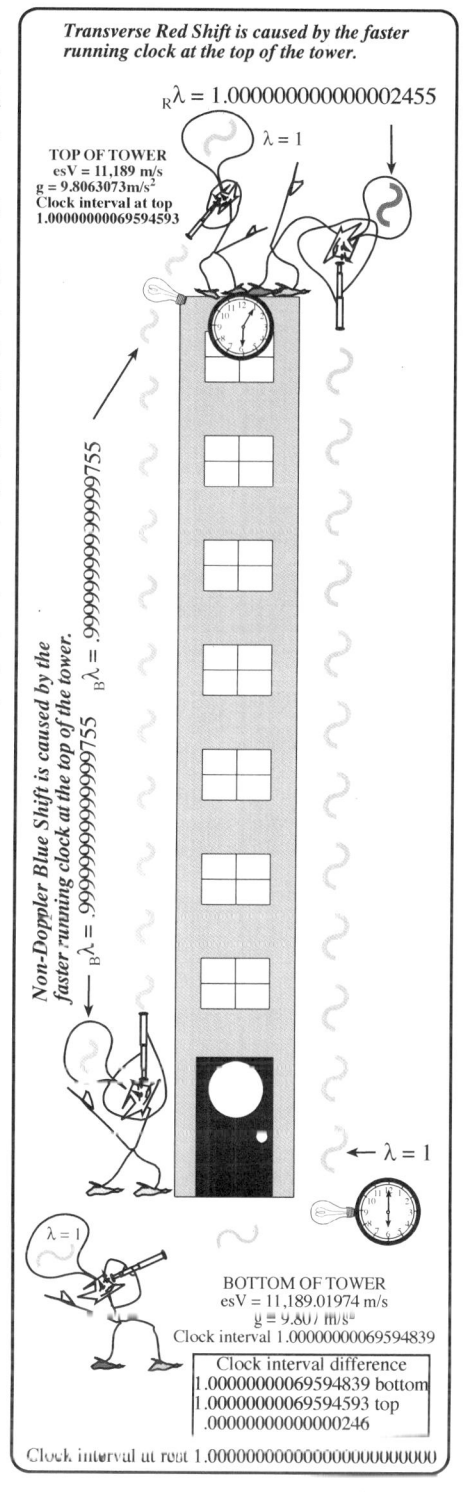

Transverse Red Shift is caused by the faster running clock at the top of the tower.

$_R\lambda = 1.000000000000002455$

$\lambda = 1$

TOP OF TOWER
esV = 11,189 m/s
g = 9.8063073 m/s^2
Clock interval at top
1.00000000069594593

$_B\lambda = .999999999999999755$

Non-Doppler Blue Shift is caused by the faster running clock at the top of the tower.

$_B\lambda = .999999999999999755$

$\leftarrow \lambda = 1$

$\lambda = 1$

BOTTOM OF TOWER
esV = 11,189.01974 m/s
g = 9.807 m/s^2
Clock interval 1.00000000069594839

Clock interval difference
1.00000000069594839 bottom
1.00000000069594593 top
.00000000000000246

Clock interval at rest 1.0000000000000000000000000

Why Einstein Was an Ignorant Fool   *James Carter*

Pound-Rebka Theories

The Pound-Rebka experiment is a simple and elegant definitive scientific measurement of the relationship between the momentum produced by force of gravity and the momentum of photons. The experiment clearly separates the relative motion and equivalent force proposed by both Newton and Einstein's gravitational force field theories from the measured momentum of photons and the measured motion and force of gravity.

The experiment uses accelerometers and linear actuators to clearly demonstrate that the gravitational motion and force of matter is in the opposite direction from the force and motion of Einstein's imaginary gravitational field theories. Motion and force are measured to be equal and opposite to the equivalent unmeasured force and motion assumed by the equivalence principle. Inertial force produces relative acceleration and gravitational force produces absolute deceleration.

The mechanical details of Pound-Rebka are quite simple. Fe-57 crystals were used as high-energy gamma photon emitters and receivers. They were placed at the top and bottom of Harvard's Jefferson Tower to emit and absorb photons of a single and very precise momentum and wavelength. As these photons moved back and forth between the top and bottom of the tower, they were carefully measured. It was found that photons received at the top were measured as red shifted and those received at the bottom were measured as blue shifted. These red and blue shifts were extremely small but precisely measured with the Mossbauer effect to be equal.

Ever since the Pound-Rebka experiment was first performed in 1959, no one has ever disputed the extreme accuracy of the results. However, a number of gravitational theorists have tried to use their own various preferred pet theories of gravity to explain the measurements in terms of gravitational fields, potential energies and other unmeasured metaphysical entities like aethers, a continuum, or spacetime dimensions.

As soon as the results of the experiment were in, many theorists claimed that it had proven both Einstein's equivalence principle and the general theory of relativity. The problem was that there were two opposing camps within general relativity proponents and both calculated the same measured values for the shifts but offered different equations and two completely different metaphysical mechanisms to explain them.

One camp imagined that higher gravitational potentials at the top of the tower caused recorded clock intervals to decrease and cause a measured red shift in the photons. The Fe-57 atom's faster clock caused un-shifted green photons from the bottom to be measured as taking a longer time to be absorbed and thus saw them as red. Gravity does not shift photon momentum and wavelength but changes the conditions under which they are measured. A transverse shifted clock emits blue photons and absorbs green photons as red. Upward gravitational velocity V_{es} at the top of the tower is less than the V_{es} at the bottom. The clock at the top has less transverse mass and runs faster from the conservation of angular momentum. The faster top clock emits blue-shifted photons with greater energy and shorter wavelengths and absorbs green photons as red shifted .

While Einstein's calculations for these shifts were verified by the Pound-Rebka experiment to fifteen orders of magnitude, the actual direction of the measured force and motion of gravity was completely ignored by his followers when they tried to explain the physical dynamics of the measured parameters. They imagined and believed that gravity and inertia were equivalent but the experiment measured gravitational shifts to be equal to inertial Doppler shifts and no equivalence was found.

Other relativity theorists also ignored gravity measurements and claimed that as photons move through the varying potentials of their imagined gravitational field, they either gained or lost momentum and energy to the field and thus changed their wavelengths while in flight. Most believed that this effect was accomplished without any change in the photon's velocity at c.

A few theorists even believe that photons slow down or speed up as they pass through their own peculiar gravitational field. Most believe photons fall in Earth's gravitational field exactly like any other material bodies. All of these theorists make calculations with their ideas that match Pound-Rebka values. Some theorists believe in the Lorentz transformation of momentum and time and some don't. Most believe in the time dilation of clocks but some don't.

Many general relativists even believe that *all* photons throughout the universe are constantly being red-shifted as they pass through expanding and curving gravitational spacetime potentials. Cosmologists use this assumption of a curving and expanding gravitational medium to explain the red-shifted Hubble photons from distant galaxies as peculiar non-motion induced Doppler effects in which photon momentum and energy are not conserved.

Hard core Einstein enthusiasts and the fools who believe in the eternal 1/1836 e/p ratio imagined the assumption of curving and expanding space justified their metaphysical belief that the 2.7°K Cosmic Blackbody Radiation can cool without violating the laws of thermodynamics. They believe that these photons were emitted when their blackbody distribution curve had a temperature of about 3000°K. The photons had a thousand times more momentum and their wavelengths were a thousand times shorter than they are today but their numbers have not changed significantly. Relativity theorists believe that more than 99% of the photon momentum and energy ever produced by the cosmos has simply disappeared down their imagined cosmic rat hole of an expanding spacetime continuum.

Relativistic cosmologists imagine the enormous amount of momentum and energy lost from this cooling process was not transferred to the rest of the cosmos, was thus not conserved. They imagined that it just disappeared into their imagined expanding spacetime continuum as if it never existed. This idea for a massless expanding dimension is in complete violation of the conservation laws for momentum and energy/mass. It is opposite the Doppler effect because these shifts in momentum and energy do not involve any change in the relative velocity between photon and observer. One major problem with relativity theorists is that they go off in theoretical tangents and are then unable to make up their minds. They seem satisfied that since all of their different ideas are equivalent they must all be correct.

Either gravity shifts a photon's momentum in flight or it does not. Quantum electrodynamics has long established that Doppler shifts only occur at the emission and absorption of photons yet there is a substantial group of Einstein's followers who believe photons can be Doppler shifted as they travel through expanding spacetime fields or a gravitational continuum.

Einstein Believed in Gravitational Waves

Einstein imagined the universe to be filled with waves of graviton particles that caused the gravitational interaction of matter. Some theorists believe that gravitons can interact with photons to change their momentum and energy. Others believe that gravitons only effect the momentum of matter and not photons. There is no direct evidence for either of these metaphysical assumptions. All experiments clearly show that all photon shifts originate at either emission or absorption and not in the medium in between, be it aether of glass.

Either the force and motion of gravity originates inside of atoms as measured or it exists as imagined and calculated varying potentials far outside the physical boundaries of atoms. It cannot be both ways. Either photons are Doppler shifted by moving atoms or they are non-Doppler shifted as they move through a stationary continuum field. Again, it cannot be both ways. Relativists have to make up their minds.

The principle of equivalence must be a great philosophical paradox to the logical and common sense minds of children who can't sense equivalence and are not yet able to comprehend its paradoxical nature. Since infancy, children have constantly felt the force of gravity pushing them upward. No sensual being would conclude anything else be it elephant, or microbe. Certainly the subconscious mind that controls a child's sense of balance would never be fooled into believing in equivalent force. How do you maintain your balance against an equivalent force?

Einstein however, was able to completely ignore his own sense of balance, at least on a conscious level, and fool himself into imagining and then believing that gravity pointed down. By using clever mathematics, he was able to fool other theoretical physicists into believing that the force of an atom's gravity of pointed down toward Earth's center from deep in the cosmos. He completely ignored the experimental physicists who were only able to measure gravity as an outward push from within Earth and who could measure no gravitational force beyond its surface.

Imagining an undetectable universal gravitational attraction caused by curving space seems to present no problems for the typical general relativity theorist. It is a standard opinion among relativists. Because the parameters can be calculated in opposite ways, they are believed to be equivalent and relative and just represent different aspects of a single gravitational continuum. What this double speak means is that because gravity can be calculated in the two opposite directions of *up gravity* and *down gravity*, the calculations are "equivalent" and thus both are thought to be intrinsically true.

Up Gravity versus *Down Gravity*

Einstein used Galileo's measurements and calculations for *up gravity* to fantasize about Newton's equivalent unmeasured *down gravity*. Einstein was able to fool his conscious mind into believing that *down gravity* was true. However, he was never able to convince his subconscious mind that *down gravity* was real because it kept compensating for and reacting to *up gravity* in order to maintain Einstein's balance while he drew equations on the blackboard. Just because the calculations for *up gravity* and *down gravity* are equivalent, does not mean they can both be true. Everyone believes in *up gravity* because they can feel it and see it happening all around them. If they didn't believe and react to it they would fall flat on their faces. To all but theoretical physicists, the four-dimensional nature of *down gravity* always remains an elusive idea that cannot quite be contemplated by a three-dimensional mind.

To believe that *up gravity* is true, you merely need to observe and measure bodies of matter undergoing the very slow mechanical process of the gravitational expansion of mass, space, and time. By contrast, if you want to imagine and then believe in *down gravity* you must open a whole can of wormholes and metaphysical assumptions. With *down gravity*, an atom's gravity does not extend from just its center to its surface, as in *up gravity*. In *down gravity* each atom has an eternal unbroken connection with every other atom in the universe. It might be easy for gullible theoretical physicists to believe in and calculate *down gravity* interactions but none have ever come up with a reasonable explanation of how it all really works. They say the ultimate "truth" is buried deep in the four-dimensional mathematics and cannot be visualized with a three-dimensional mind.

The experimental truth of *up gravity* is far easier to visualize than the virtual truth of *down gravity's* n-dimensional math. While Einstein's calculations may perfectly match measurements, his equations contain non-physical parameters like waves in potential fields that only exist in the predictions of his metaphysical ideas. Einstein was a fool to think he could take the *up gravity* measurements of experimental physicists and then turn them upside down and inside out to create a *down gravity* theory that is perfectly equivalent to *up gravity* values. Certainly his counter-intuitive, mathematical theory of *down gravity* works beautifully, but why would he bother. The real mystery here is why he chose to abandon the *up gravity* he had intimately experienced his whole life.

Gravitational Expansion is a Principle and Not a Theory
The principle of gravitational measurement calculates the values of its parameters and needs no theory to explain it. Gravitational expansion is simply what we feel gravity to be and the measurement of what gravity does. You only need a theory of gravity to explain an opposite downward gravitational force field like Newtonian's universal gravitation or Einstein's general relativity.

The gravitational expansion of mass, space and time is not even an assumption. It is just calculations based on physical measurements of $F = ma*d$. Force equals mass times a combination of absolute acceleration and deceleration. This is not a theory of gravity, but rather just the substantive values of the physical measurements of mass, space, time and gravity.

Falling bodies do not change their state of motion until they are hit by Earth's upwardly decelerating surface. Photons do not change the state of their motion or momentum until they are reflected or absorbed by atoms. The gravitational expansion of Earth is measured to be a simple local mechanical process of constant upward force and momentum.

Gravity is Child's Play
While Einstein's theory of equivalent gravitational force can only be fully understood through complex and infinite mathematical equations, the gravitational expansion of mass, space and time is a simple, local, and mechanical process that can be explained to and completely understood by the smallest child. Ask a child to repeatedly jump off a chair, watch the floor rushing up, feel it strike them and then continue pushing them upwards. The simple upward force and motion of gravity will quickly become apparent to a child because he or she have both felt it and seen it happening their whole lives. The gravity that children feel is easily measured, calculated, explained and understood. By contrast, no child or even Einstein himself could completely understand and visualize the workings and mechanics of an opposite senseless and unmeasured universal multi-dimensional spacetime continuum field.

Equivalent Momentum as a Cause
Einstein imagined that physical momentum continually flows out from the equivalent momentum of the gravitational field and into falling bodies to increase their downward velocity.

When Galileo first made experimental measurements of gravity, he may even have understood the true upward motion and force of gravity. Certainly, his dropping of different weights from the top of the Tower of Pisa would indicate upward gravitational motion of Earth's surface. Then just a year after Galileo died, Newton was born and began inventing equations and calculations to turn Galileo's measurements upside down. Many years later, Einstein imagined his version of the equivalence principle and invented equations to turn gravity inside out and backwards with his calculations of a curving gravitational continuum field that altered the scientific method's measurement principle by reversing the temporal direction for cause and effect. Einstein offered no mechanism of how the physical effect of momentum could just appear from apparently empty space without an initial physical cause other than an undetectable equivalent force caused by curving spacetime.

Planck's Constant

This mechanical explanation of the Pound-Rebka's gravity and photon measurements is not meant to be some new theory of gravitational force and motion. It is strictly a measurement of the gravitational expansion of mass, space and time using photon electrodynamics and the Newtonian laws of force and motion. This is a simple description of the Pound-Rebka apparatus used to measure mechanical force and motion and no "theory" of gravity is involved.

There are only three mathematical equations used in this explanation. One for the force and momentum of matter $F = ma*d$, Newtonian Force is equal to mass times a combination of absolute acceleration and deceleration. One for the Lorentz transformation of mass and time $\sqrt{1-v^2/c^2}$ and one for the constant mass and wavelength of photons $h = m\lambda c$. Planck's constant h is equal to the photon masslength constant $m\lambda$ times the speed of light c. The value Planck's constant $h = 6.6260755 \times 10^{-34}$ is the mass of a photon in kilograms with a wavelength of one meter or the wavelength of a photon with a mass of one kilogram.

The Nature of Photon Momentum

The only aspect of the photon measured in the Pound-Rebka experiment is its momentum $p = mc+/-v$. All theories of the photon use momentum as its primary parameter of measurement, as do all theories of matter consider momentum as the primary parameter in which measurements of the motion of mass are made.

This explanation of Pound-Rebka is not meant in any way to be a new theory of photons. The photon model used in this description is strictly a generic photon made up from simple photon measurements of momentum, wavelength, velocity, and energy. Regardless of what your theory of the photon might be, you must use these same measured values of momentum to quantify your version of the photon. A photon's wavelength and energy are determined from measurements of its momentum. It makes no difference whether you believe the photon to be a massless wave of "pure energy" or a particle of mass with the kinetic energy of its motions. Whether or not the photon has mass is irrelevant to this experiment. Any body's rest mass is a calculated quantity that is only implied through the measurement of its momentum. Momentum is a relative quantity and we can never measure a photon's exact mass or energy because we cannot separate the Doppler effects of its absolute momentum of mc from its relative momentum of mv.

The sole measurement made in the Pound-Rebka experiment is the relationship between the relative momentum of photons and the changing absolute momentum of gravitational escape/surface velocity V_{es}. Photon energies and wavelengths can be calculated from these results but they are secondary photon parameters and not part of this experiment that only measures momentum. The Doppler shift of photons is purely a measurement of momentum.

The photon model used in this experiment is simply a moving point of momentum p = mc. All photon theories begin with this measured point of momentum. How or why these points move or what kind of medium they move through has no bearing on how we measure a photon's momentum. In my illustration of the experiment, I show the photons as *masslengths* of cosmic string with wave-like motion. While these illustrations may represent some aspects of truth, the energy and wavelengths of photons are not used to explain the Pound-Rebka results.

Measuring Relative Motion in Space Travel

If we could travel in a photon powered rocket at one-third the speed of light, the photons emitted and absorbed by atoms in the direction of our motion would have twice the momentum and half the wavelengths of the photons absorbed and emitted against the direction of the rocket's motion.

However, within your space capsule, these photons would be measured to have the same momentum p = 1 and wavelengths $\lambda = 1$ as they had when the capsule was at rest on Earth's surface. Even though the photons emitted and absorbed within the capsule are highly red and blue shifted with both transverse and direct Doppler shifts, these shifts cancel each other and cannot be detected with measurements inside the capsule. Photons emitted and absorbed within the capsule are measured to have an un-shifted wavelength of $\lambda = 1$.

If we could travel much faster to very near the speed of light, we would measure the forward facing 2.7°K CBR microwave photons to have a very high temperature with the momentum of X-ray photons and gamma photons. By contrast, the backward facing 2.7°K CBR photons would have the momentum of radio wave photons and a temperature of very near to absolute zero.

All calculations of a photon's mass, wavelength, energy and frequency are based on the measurement of its momentum. Only at absolute photon rest (located at a velocity of 375 km/s in the direction of Aquarius) would we be able to measure a photon's true non-Doppler shifted momentum, wavelength, and mass. At such a point within zero momentum frame of photon rest, all of the CBR photons from all directions would have the same blackbody wavelength distribution curve and the same overall temperature of 2.726°K.

Transverse Doppler Shifts in Momentum and Time

All randomly moving bodies such a clocks, bullets and photons possess a single absolute momentum mv vector relative to the stationary space of photon rest. As a body's momentum is increased from zero at photon rest, it accumulates kinetic energy $e = mv^2/2$. As each body's individual momentum is increased from zero at rest, its energy/mass constant increases with velocity $e/m = v^2/2$ in direct proportion to momentum. A body's energy/mass increases and decreases with each acceleration and deceleration of its absolute momentum vector. At any position within photon rest, an atom's mass is at its minimum value of one $m = 1$, its momentum is zero $p = mv = 0$ and its photon clock has the maximum rate of one $t = \lambda/c = 1$ and emits a $\lambda = 1$ photon. Any accelerated motion away from this position of zero momentum rest requires kinetic energy and increases an atom's energy/mass with the added momentum but does not change its angular momentum. As a result, the increased mass slows the atom's clock interval to $t = 1+$ and it emits a $\lambda = 1+$ photon. As a body's absolute velocity vector is increased, its mass increases $m/\sqrt{1-v^2/c^2}$ to greater than one $m => 1$ and the rate of its photon clock slows to intervals $t/\sqrt{1-v^2/c^2}$ of greater than one $\lambda/c => 1$.

Lorentz Transformation of Mass and Time

When a clock is accelerated to a high velocity, its mass is increased and the length of its recorded intervals increase by an equal amount and when it is decelerated, its mass decreases and it ticks faster. This effect is a result of the conservation of angular momentum. In the case of a bullet, its energy is the combination of the momentum of its linear path and the angular momentum of its rifled spin. The bullet's total measured energy combines the relative kinetic energy of its velocity $e = mv^2/2$ and the absolute rotational kinetic energy $e = I\omega^2/2$ of its spin. A photon's total energy is like a bullet except that the relative energy of its momentum is equal to the absolute spin energy of its angular momentum: $e = mc^2/2 + mC^2/2 = mcC$. Doppler effects shift a photon's linear relative energy but not the absolute energy of its spin.

In the direct Doppler effect, the photon's momentum is increased or de creased in direct proportion to the direction of the emitting atom's absolute momentum vector. Photons emitted or absorbed with the atom's direction of motion will be blue-shifted and photons emitted or absorbed against its motion will be red shifted. Photon paths at right angles to its vector of motion will have no direct Doppler shifts but will undergo transverse Doppler effects. The transverse Doppler effect is the same in all directions and it red shifts emitted photons and blue shifts absorbed photons.

In a transverse Doppler effect $\sqrt{1-v^2/c^2}$, an emitted photon's momentum is decreased in direct proportion to increases in the atom's linear momentum. The increased momentum of an accelerated atom does not change its conserved angular momentum $I\omega = mvr = $ time. When a rotating body's mass is increased and its radius remains constant, its rate of rotation must decrease. This increases intervals of the atom's clock and it slows. The time dilation of moving clocks is caused by the conservation of angular momentum as the atom's linear momentum is increased.

Theoretical physicists claim the equivalence principle has been measured to be correct many times and to a precision of many orders of magnitude. However, all of these confirmations have been with null results. An example of such an experiment would be to place a very accurate accelerometer on a falling body within a vacuum. When it is determined that the accelerometer registers no downward change in motion, the claim is made that the equivalence principle had been proven to the limits of the accelerometer's accuracy. The Pound-Rebka experiment has long been claimed to be a proof of the equivalence principle but here again the verification is in the form of null results. All failures to detect a downward force and motion of gravity are seen as experimental proofs of equivalence. The equivalence principle basically says that "gravity cannot be directly measured". Whenever an experimenter fails in an attempt to measure gravity, relativists declare it as a "proof" of Einstein's equivalence principle.

The Mossbauer Effect
The Mossbauer effect is the phenomena that Pound and Rebka used to make their very accurate measurements separating transverse and direct Doppler shifts.

Atoms emit photons that travel an unlimited distance at c to other atoms that either absorb or reflect them. Atoms reflect most photons and can only absorb a limited number of wavelengths. This number decreases as the momentum and energy of the photons increase. The greater the photon's momentum and energy, the less likely it can be absorbed and not reflected by an atom. In the longer wavelengths, like visible and radio photons, a very wide range of photons can be absorbed and emitted. In the shorter wavelengths like X-rays and gamma photons only a very limited number of wavelengths can be emitted and absorbed by a particular atom.

As the momentum and energy of photons increase, atoms can absorb fewer and fewer photons until finally they can only emit and absorb a single photon of a precise wavelength. In the case of the Pound-Rebka measurements, the atom used is Fe-57 and the photon is a gamma ray photon that has a momentum, energy and wavelength of exactly one $\lambda = 1$. An Fe-57 atom at rest can only emit or absorb this gamma photon. The Fe-57 atom reflects all similar sized photons and is only able to emit or absorb this single exact photon. When this photon travels from the bottom of the tower to the top, it maintains its wavelength of $\lambda = 1$.

When two of the Fe-57 crystals used in the Pound-Rebka experiment are placed apart horizontally on level ground, they are each able to readily absorb the $\lambda = 1$ photons emitted by each other. However, if either crystal is put into even a slight amount linear motion, all absorption of $\lambda = 1$ photons stops. It doesn't matter which crystal is moving. Even the tiniest amount of relative motion between the crystals produces direct Doppler shifts that prevent the photons from being absorbed.

When Pound and Rebka placed two Fe-57 crystals apart vertically in the Jefferson Tower, there was no photon absorption at either the top or bottom crystal even though they appeared to be at rest and maintained an exact vertical distance of 22.5 meters. There are only three possible ways to explain this effect. Either the photons acquired Doppler shifts instantly at emission or absorption or they acquired the shifts gradually as they traveled from emitter to receiver.

The receiving and emitting clocks at the top of the tower are moving at a lower escape/surface velocity V_{es} by than the clocks at the bottom. This difference in absolute gravitational momentum causes clocks at the top of the tower to be less massive and have shorter intervals than clocks at the bottom. This slower upward gravitational motion has red-shifted the receiver to a wavelength of $\lambda = 1.0000000000000025$. The $\lambda - 1$ green photon from the bottom can no longer be absorbed by an Fe-57 atom at the top of the tower that can only see red.

Linear Actuator Measurements

A simple measurement with an accelerometer reveals the true value of the motion producing the shift. Both crystals are found to be decelerating upward with the force of gravity.

In the time that the photons take to move between crystals, the top of the tower's V_{es} upward velocity has decreased by $V = 1 = 7.36 \times 10^{-7}$ m/s and it is this decrease in the receiver's gravitational momentum that produces the red transverse shift that prevents the green photons from being absorbed and causes them to be reflected. This measurement shows that even though the distance between them appears to remain constant, the top and the bottom of the tower are moving together with gravitational deceleration and apart with gravitational expansion. The value of this absolute gravitational motion can be calculated with simple physical measurements of force and motion and then the Mossbauer device can be used to confirm the absolute values of this motion. These measurements separate the transverse shifts are caused by absolute motion and the direct Doppler shifts are caused by relative motion.

There is no logical, philosophical or experimental reason to believe Einstein's mysterious unmeasured force field would cause the photons to shift their momentum in flight. People like Einstein who fool themselves into believing in imaginary gravity theories they can't even feel or measure do so for peculiar psychological, emotional, and spiritual reasons that have nothing to do with experimental physics.

In order to make the transverse red shifted Fe-57 atoms at the top of the tower absorb the photons from un-shifted atoms at the bottom, it is necessary to put the bottom emitter on a very slow linear actuator that can move upward at the precise velocity of V = 1. As the actuator speeds up to this velocity, it direct Doppler shifts the photons to blue by the same amount as the receiver's transverse red-shift. When the shifted actuator blue photons are absorbed by the red shifted receiver they are measured as green.

In the case of photons moving from the top emitter to the bottom receiver, the linear actuator is set to move downward at V = 1. The blue transverse shifted photons from the top emitter are measured as green by the direct Doppler red shifted bottom receiver moving down at V = 1.

In the first case, direct Doppler blue shifted photons are canceled to green by a transverse red shifted receiver and in the second case, transverse blue shifted photons are measured as green by a direct Doppler red shifted receiver. Both of these shifts occur from measured absolute linear velocity. The direct shifts are caused by the relative velocity of V = 1 and the transverse shifts are caused by the differences in escape/surface velocity between the top and bottom of the tower. There is no logical reason to believe, as Einstein did, that the transverse shifts are caused by imaginary unmeasured gravitational potentials and fields. If he really believed this was true, why didn't he explain how the measured accelerometer readings do not produce shifts.

Principle of the Gravitational Expansion of Mass, Space, and Time

When we measure and calculate the physical mechanics of gravitational motion and force with Newtonian accelerometers and cesium clocks we logically and philosophically arrive at the gravitational expansion of mass, space and time.

Gravitational expansion is a physical principle of measurement and not a metaphysical theory of gravity because it makes no assumptions other than the accuracy of accelerometers and $F = ma*d$. Force is equal to mass times an unknown combination of deceleration and acceleration. This is the only equation necessary to explain the physical mechanics of the principle of gravitational expansion. Any "theory" of gravity must present metaphysical assumptions to explain why the physical measurements of upward gravitational motion and force are false. Einstein's attempt at falsifying experimental measurement was his theory of equivalence.

Gravitational expansion is a principle of measurement. Its how we feel the force of gravity with our bodies and what gravity does and looks like when we measure it. Skydivers in free-fall feel motionless as they watch Earth rush up to meet them. This is a simple common sense observations that employ all their senses. Skydivers have no sensory input that would lead them to believe they were being accelerated towards Earth. They can easily feel the air pushing them upward as Earth's surface rushes toward them.

For skydivers to invent a "theory" of gravity to counter the upward principle of gravity based on their own sensory observations, they must first psychologically overcome their sensual feelings of watching a rising Earth. To defeat these feelings, skydivers must try to imagine they are somehow being unfeelingly pulled toward a stationary Earth by an otherwise undetectable force field. To do this, they can imagine different ideas like gravitons, aethers, dimensions, continuum fields and spacetime. They can use one or a combination of these intangible mathematical ideas to create a mechanism that will invalidate the workings of gravity they see and feel as they float above Earth on a rising column of air. This mechanism would be used to invalidate the measurements of Newtonian accelerometers and Cesium-133 clocks.

Philosophy of Gravity

The philosophical problem with all gravitational field and particle theories is that no gravitational fields or particles have ever been detected and measured and the physical problem with these theories is that they completely ignore the actual experimental measurements of gravitational motion, force, and time that are made with accelerometers and clocks.

The metaphysical assumptions and mathematical calculations behind the Einstein apologists' explanation of the Pound-Rebka duplicate the measured values of the experiment perfectly. The only problem is that the math incorporates the unmeasured force and motion and equivalent momenta of both electromagnetic and gravitational field potentials. Einstein imagined gravity and inertia to be equivalent but not equal. He believed that the measured inertial force and motion of electromagnetic photons was always equivalent to the unmeasured motion and force of gravity. Einstein's purpose in inventing the equivalence principle was an attempt to invalidate Galileo's measurements of gravitational force and motion as demonstrated by Newton's pendulum clock. Even though the pendulum's motion is obviously a pure measurement of upward acceleration, Einstein imagined that gravity was an unmeasured downward force caused by the curving of an unseen substance called a four-dimensional spacetime continuum. However, gravity is always measured as a four-dimensional absolute motion of mass and time through the void of space.

The essential difference between general relativity and gravitational expansion is that with relativity, gravity is caused by the imagined curvature of the spacetime surrounding matter and photons and with gravitational expansion, it is the mass-time within atoms and photons that is measured to actually curve.

Both general relativity and gravitational expansion explain gravity with opposite four-dimensional geometries of curving space and time. In the first idea, it is Einstein's imagined dimension of spacetime that curves inward and with gravitational expansion it is the actual measured dimensions of atoms that curve outward.

If the upward force of gravity doesn't produce real upward inertial motion, how is it that the measured inertial motion of the linear actuators is able of to Doppler shift the gravitational motion of photons? Pound-Rebka clearly shows that the inertial motions of both photons and gravity are one in the same equals and not equivalent opposites.

Gravity Comes from the Inside of Atoms and Not the Outside

The motion and force of gravity come completely from within the interior of the atom's mass, space, and time and not from other atoms in the universe at large. The upward force of gravity extends only to the outer surface of each photon, atom, moon, planet, star, and other solid bodies of matter.

While the rates clocks are a measure of transverse shifts, it is changes in momentum and not changes in time that is the true physical parameter of the transverse shift. Time is just an imaginary parameter to quantify momentum $p = ms/t$ and like the idea of space has no separate physical existence. Space and time are just the imaginary handles for the measurement of mass. Photon momentum $p = mc$ and angular momentum $m\lambda c/2\pi$ are the two separate configurations of mass, space and time.

Protons and electrons are measured as individual bodies of circlon shaped mass. Each projects a circlon shaped charge chain "field" that pulls them together, or pushes them apart. All of their interactions are primary transfers of momentum and secondary transfers of energy/mass. Momentum and energy are not separate things. Momentum is the primary entity and energy is the interaction between momentum and force. Whereas momentum is absolute, eternal, and conserved, energy is relative and momentary. Total energy is conserved but the conservation law for energy is secondary to the primary momentum conservation law. Moving bodies with identical conserved momenta can have greatly different values of conserved kinetic energy. The cannonball has far more energy than the cannon but their momenta is the same. A single force measured as energy $e = mv^2$ is divided into two equal momenta $mv = mv$ that produce two unequal energies $e = mv^2/2 + e = mv^2/2$.

A moving body with a momentum of one $p = 1$, a mass of one and a velocity of one has a kinetic energy of one-half $e = mv^2/2$. Another body with a momentum of one can have a mass of 1/10, a velocity of 10, and an energy of 5. A heavier $p = 1$ body with a mass of ten and a velocity of 1/10 would have an energy of 1/20.

Momentum is the constant motion of mass and energy is the measure of any change in that motion. Energy provides the force to create momentum and change in momentum produces the force necessary to create energy. Energy is the relative, positive or negative measure of a body's change in momentum. Energy is absorbed when a body is accelerated and energy is produced when a body is decelerated.

The gravitational expansion of mass, space and time might be called a "dynamic dimension" but it is not in any way a field, an aether, a spacetime continuum or anything else that would give substance to space. There is no theory required for gravitational expansion because gravity is just a physical measurement and physical principles of measurement cannot be theories. There is no question as to what gravity does experimentally. Gravitational theorists try to believe in a field that cannot be measured. Relativity theorists all claim that the equivalent force and motion of gravity point down yet the physical results of Pound Rebka and all other accelerometer measurements show conclusively that gravitational force and motion point up, not down.

Photon, Transverse and Gravitational Times

The standard for inertial photon time is the one-dimensional conservation of momentum. The natural interval of photon time is the light year. The standard for inertial transverse time is the two-dimensional conservation of angular momentum. The natural interval of transverse time is the angular momentum of the photon $h/2\pi = m\lambda c/2\pi$. The standard for gravitation time is the constant upward three-dimensional V_{es} force and momentum of Earth's surface. The natural interval of gravitational time is the year.

The natural interval of Earth's gravitational time clock is one year and the natural interval of Earth's transverse rotational time is the day. Common forms of gravitational clocks like pendulum clocks do not produce pure intervals of gravitational time because on Earth, pendulum clock intervals are changed at different latitudes by the opposite transverse time of the centripetal forces produced by Earth's inertial rotation. Only at the poles do pendulum clocks produce pure gravitational time intervals.

The synchronization of a pendulum clock and an atomic clock can usually only be accomplished at two or three orders of magnitude but the Pound-Rebka measurement is a unique example where the two time flows can be synchronized to within fifteen orders of magnitude 1.00000000000000025.

Gravitational time and photon time flow in different directions. The flow of gravitational time is up and the flow of inertial time is down. The upward force of gravity that drives the speed of the pendulum is counteracted by Earth's downward centripetal force that slows the pendulum. Earth's upward force is gravitational time and its downward centripetal force is inertial time. These two times merge into the recorded intervals of a pendulum clock.

The intervals of transverse rotational time are the angular momentum of the photon and the angular momentum of the electron $I\omega = m\lambda C/2\pi = e_m a_o \alpha C$. The recorded intervals of all clocks are based on the conservation of angular momentum. In atomic clocks it is the internal rotations of the atom that is the standard of time. When an atom is accelerated, its momentum and mass increase but its angular momentum remains constant. This causes the rotations within the atom to slow and the clock's recorded intervals to become longer.

Acceleration and Deceleration of Gravitational Time

The force we measure with an accelerometer is usually an unknown combination of absolute acceleration and deceleration. $F = ma*d$. With the Pound-Rebka experiment, we are able use gamma photons to precisely measure the difference between the acceleration and deceleration of gravity.

Einstein imagined gravity as an unmeasured force that produces virtual acceleration of mass within a four-dimensional spatial force field surrounding all bodies of mass. This is a metaphysical description because it completely ignores the actual physical measurements of gravitational force and motion. It is based on equivalent acceleration that produces none of the changes in motion that we measure with accelerometers. What Newton and Einstein failed to understand was that there is no way for an accelerometer to distinguish between absolute acceleration and deceleration and that their equivalent acceleration of gravity was actually measured to be deceleration. Gravity is an acceleration/deceleration dichotomy that is equal and opposite but not equivalent.

The problem with both special and general relativity is that they never embraced deceleration as being a real and distinct equal to acceleration. If all linear motion is assumed to be relative, there can be is no physical distinction between acceleration and deceleration. The contradiction that exists within both Einstein's theories is that the Lorentz transformation and Pound-rebka provide experiment measurements of absolute linear motion. Many experiments have measured that linear acceleration causes mass to increase and clocks to slow while deceleration causes mass to decrease and clocks to speed up.

The Pound-Rebka experiment makes very precise Doppler shift measurements of the momentum transfers between matter and photons produced by the combined complimentary acceleration and deceleration of the gravitational expansion of mass, space and time. These experimental measurements of momentum transfer show the clear distinction between gravity's upward accelerated motion and the equal and opposite deceleration of Earth's surface/escape velocity.

A simple example of acceleration and deceleration being produced by changing force and motion is the centripetal force produced by a rotating body. Unlike so-called radial centrifugal "force" that is imaginary and without any measurable parameters, centripetal force is an actual force that can always be measured. Centripetal force is measured with an accelerometer as a moving force F = ma directed toward a body's axis of rotation. Rotating bodies are equally accelerated and decelerated at the same time by centripetal force. It is analogous to gravitational force in that both are equal combinations acceleration and deceleration. The difference is that the direction of centripetal force is towards a body's center and the direction of gravitational force is away from its center. Centripetal force is an inward two-dimensional pull and gravitational force is an outward three-dimensional push. Inertial force is one-dimensional on a single vector, centripetal force is a two-dimensional inward force on circular plane and gravitational force is a three-dimensional outward force between a body's center of mass and an infinite number of directions in space.

An example of the deceleration of gravity is a rocket engine in a fast low Earth orbit such as a space shuttle orbit. For the rocket to move a satellite to a much higher orbit such as a GPS orbit, it must decelerate it to both a lower orbital velocity and a lower escape/surface velocity V_{es}. The positive energy of the rocket fuel is used to decelerate the satellite and remove its kinetic energy as it slows to lower escape/surface V_{es} and orbital velocities V_o. This example is experimentally verified by atomic clocks onboard GPS satellites that increase their rates when they are decelerated into higher and higher orbits with less and less escape/surface velocities and orbital velocities.

Earth Falls Up

The Pound-Rebka experiment provides definitive proof that Einstein was wrong about both the direction of gravitational motion and force and the relativity of photon motion. The Pound-Rebka values are not the result of some otherwise hidden transformation of space and time. The measured results are a combination of transverse and direct Doppler shifts produced by measured changes in actual inertial motion.

Mathematics can never be used as proof of a theory of physical measurement. Physics is the proof and mathematics is just a tool used to quantify that truth. The metaphysical assumption is a tool to mathematically counter the results of experimental measurements. The only mathematics needed by Pound-Rebka to quantify gravity and photons are $F = ma*d$, $p = mc$, and $h = m\lambda c$.

The true nature of gravity is a dichotomy between the equal and opposite inertial acceleration of matter and the deceleration of gravitational momentum and time. What is usually referred to as the "acceleration" of gravity is, in reality, always measured as a balance between both deceleration and acceleration.

The Harvard Fantasy of Equivalent and Relative Motion

Harvard physicists often tout Pound-Rebka as an actual measurement of the hidden gravitational potential produced by the curvature of gravitational space that is calculated and predicted by general relativity's equivalence principle. However, the use of actuators to produce direct Doppler shifts to cancel the transverse shifts of gravitational motion proves quite conclusively that there can be neither a gravitational potential field nor a curved gravitational spacetime continuum. No gravitational potential is required for the measured direct shifts and if such a "potential" were actually to exist, it would change the measured values.

The only space that curves is not the four-dimensional external spacetime continuum space surrounding atoms but rather the three-dimensional internal gravitational space defining the shape and intrinsic inertial expansion of atoms. The gravitational expansion of mass, space and time is the true measure of gravitational motion and force and it extends only to the surface of each atom. No measurement has ever been made of an occult gravitational attraction between atoms resulting from an all-pervasive gravitational potential field or from a curvature of spacetime. All such fanciful ideas may be psychologically and emotionally pleasing fantasies for Harvard theoretical physicists but they are definitely not philosophically pleasing for Harvard's experimental physicists. In reality, these ideas are nothing more than the purest of metaphysical speculation. The simple true nature and cause of gravity revealed by the Pound-Rebka experiment is that Earth's surface falls up with a constant upward velocity that is accelerating and decelerating at the same time.

The Common Sense Principle of Cause and Effect

The basic physical components of atoms and photons are mass, space, time and gravity. Their measurable physical parameters are force, acceleration, momentum, wavelength, and energy are calculated with scientific method's common sense principle of cause and effect.

The terms physical, metaphysical, assumption, principle and theory are used in several different combinations throughout this book and a proper understanding of their meanings is essential for the comprehension of the ideas presented herein. The philosophy of physics combines the principle of physical measurement with the metaphysical assumptions of imagination to create theories to define reality. Philosophy is the structural relationship between the common sense measurements of physical parameters and the non-sensory and mathematical interpretations of these measurements in terms of metaphysical assumptions that cannot be measured or do not obey the principle of cause and effect.

Cause and Effect of Gravity and Photon Momentum

The most important principle of physical measurement is the temporal direction of cause and effect. The direction in time of Newtonian force and motion runs counter to the temporal direction of cause and effect in Einstein's equivalence principle.

The Pound-Rebka experiment establishes that change in momentum is both the cause and effect of force. Momentum's direction in time is always measured from cause to effect. In the equivalence principle, the directions in time for cause and effect are turned upside down. Einstein's equivalent gravitational force becomes the cause and downward equivalent gravitational motion becomes its effect. The Pound-Rebka experiment does not detect or measure either equivalent force or downward motion.

Active force changes timeless momentum into energy. <u>Change in momentum is the cause and effect of force and energy is divided between the two momenta.</u> This is a simple cause and effect temporal principle of measurement and not a theory. A metaphysical theory would say that force can also be a cause when it constantly appears from within a continuum to make the momentum of falling bodies into an effect.

In his general relativity theory, Einstein imagined reversing cause and effect so that the force of gravity causes the increasing momentum of falling bodies. This transformation in direction from *cause and effect* to *effect and cause* is nothing short of trying reverse time itself. *Momentum>force>energy* is the physically measured direction of time for cause and effect. In Einstein's upside down idea of *energy>force>momentum*, energy from space causes equivalent force that produces momentum as an effect. This idea creates contrary non-experimental definitions of the concepts and calculations of *momentum, force, and energy*. Certainly, Einstein was able to imagine and calculate his reversal of physical time but he was never was able to find any experimental evidence of gravitational energy coming out of space. All measurements show gravitational energy to be the result changes in gravitational momentum at Earth's surface.

Pure Energy is an Unreal Imaginary Idea

Standard model physics and Big Bang theorists all use Einstein's upside down and backwards idea of "pure energy" as prime mover in physical dynamics. They all want to believe the universe began as a singularity containing only pure energy and no momentum. In experimental physics, there has never been any evidence of pure energy causing anything and there is no reason to believe in any such thing as pure energy. Energy is always the measured result of momentum and the force that changes momentum. Momentum and force are the primary parameters of the motion of atoms and energy is just the secondary idea used to quantify their relationship. Energy is just a momentary measurement and does not exist beyond a change in momentum. A moving body contains momentum but does not contain energy. Energy is only produced or absorbed when momentum is changed by a force. All values for energy are derived from measurements of momentum/force.

A body's absolute momentum is measured as energy/mass $e/m = v^2/c$ This value increases from zero to infinity as a moving body approaches the speed of light. The energy/mass of a photon is $e/m = cC$

$p = mv$
$F = ms/t^2$
$e = mv^2$
$F = mv + mv$

$^oE = mv^2/2 + mv^2/2$
Photon $e/m = cC$
Atomic $e/m = CC$
Momentum $e/m = v^2 - c^2/c$

Momentum, Force, and Energy

Momentum is the motion of mass through absolute space and time. $p = ms/t$
Relative Momentum is mass in lineal motion.

Absolute Momentum is mass in rotational motion.
A quantity force always bifurcates into two equal momenta.
A force always produces into two unequal energies.
Force is a momentary change in momentum produced by momentum.
Momentum is eternal and a force is an individual interval of time.
Energy is the idea used to quantify a unit of momentum/force.
Energy is not a physical entity like momentum/force.
Energy is a duality calculation of a momentum/force reaction.
Relative Energy is a change in linear momentum.
Absolute Energy is a change in angular momentum.
Energy is the measure of a momentum/force event.
Energy is the dual effect of two colliding momenta.
The collision force changes two equal quantities of momentum.
Each new equal momenta has an unequal quantities of energy.
Momentum and energy are cause and effect.
Einstein got it backwards. He imagined that energy was an independent absolute substance contained in spacetime that could cause force and change momentum. The truth is that the motion of energy/mass (momentum) is the substance of reality and energy is just a half measure of its change.

All measurements of energy contain differing portions *relative energy* and *absolute energy*. Photons have exactly equal quantities of *relative energy* and *absolute energy*. They all move at c on one dimensional vectors and their momentum is calculated to have relative energy. Photons also spin at C and their angular momentum is calculated to have absolute energy. Two moving atoms have the relative energy of their linear velocity and the absolute energy of their internal rotational velocity.

Momentum and Force are the Primary Parameters of Reality

Momentum and force are measurable entities. They are individual quantities of mass, space, time, and gravity. Momentum and force are the primary parameters of existence and energy is the secondary concept of their measurement. Energy is the relationship between momentum and force and has no independent existence of its own. Energy is the interaction between force and momentum and cannot exist independently of momentum or be measured independently of force. Force and energy are always relative to momentum and momentum is always absolute to the Zero Momentum Photon Frame. All photons move at c relative to the ZMPF. This is not a "preferred" rest frame. ZMPF is measured to be the *only* possible stationary frame. All other frames move relative to ZMPF and are subject to the direct Doppler shifts and Lorentz transformations caused by inertial motion. Even though the ZMPF is just an imaginary void of space, its absolute location can be accurately measured by finding a point of rest -c for the photons of the 2.726°K Cosmic Blackbody Radiation. This point has been measured to move relative to Earth at about 375 km/s in the general direction of Aquarius.

Cause and effect defines the temporal direction of time for both gravitational time and photon time. There is no question as to the direction of cause and effect time in photon interactions in which a photon's momentum is the cause that always precedes the effect of a photon's force and energy. Momentum is the primary absolute quantity of cause and force is the secondary effect of a change momentum that is quantified as energy. Momentum is an unmeasured passive cause and force is the active effect for the measurement of energy and mass. Force alters both the quantity of a body's momentum and the direction of its vector of motion. Momentum is necessary to produce force. Force is an absolute quantity and can only be measured as a change in momentum.

General relativity proposes the cause of gravity to be a momentum-less point of mass located at Earth's center that projects momentum causing downward force to all bodies on Earth and in the surrounding universe. In aether and field theories like general relativity, the temporal direction of cause and effect is reversed. Gravitational energy and force are imagined to pop out of the gravitational field and cause the effect of acceleration of bodies toward Earth. In the gravitational expansion principle, upward momentum is the cause of the effect of upward gravitational force. Gravitational energy comes up from Earth's surface and not down from the sky.

The Common Sense Direction of Time
The common sense directions of cause and effect are imagined to be reversed in Einstein's non-sensual field theory of gravitational mass, space, and time. There is no logical, philosophical or physical reason to believe that energy and force can ever precede momentum in any cause and effect interaction.

In general relativity, unmeasured force is assumed to be the cause of the unmeasured relative momentum of falling bodies. In such field theories, inertial photon time and gravitational time move in opposite directions for purposes of cause and effect. This is the backwards Newtonian cause and effect of $F = ma$. The cause and effect direction of Newton's formula is $ma = F$. The effect of force can never precedes the cause of accelerated or decelerated momentum as it does in relativity theories where unmeasured force is imagined to create the equivalent momentum of falling bodies.

The Pound-Rebka measurements establish momentum as the cause of both the forces of gravity and photons. There is no evidence for an unmeasured downward force of gravity being the cause of unmeasured downward momentum. In the principle of the gravitational expansion of mass, space and time, upward gravitational momentum $p = mV_{es}$ causes the measured effect of upward gravitational force. In both the principle of gravitational expansion and the principle of photon dynamics, change in momentum is always the cause of force and energy effects.

When a stationary "falling" body impacts the ground, force is produced by Earth's upward momentum. The force is quantified as energy of the "fall". When a photon is absorbed by an atom, its momentum is added to the atom by a force that is measured as an increase in the atom's kinetic energy. The mass of the photon is added to the atom as an equal part of the energy. Mass, momentum, and energy are all conserved independently. Energy/mass is the single conserved component of atoms and photons $e/m = c^2$.

Both momentum and force are dynamic entities and energy is the conserved quantity of an individual momentum/force interaction. Momentum and force are principles of measurement and energy is their value. Einstein tried to turn this relationship around by imagining that "pure energy" was the primary autonomous substance of existence. He Believed energy itself that could produce force, create momentum, and even transform into the mass of atoms. No experimental physicists have ever measured Einstein's pure energy transforming into force and momentum. Newton understood the primary physics of force and momentum and did not include energy in his measurements and calculations. Energy is not a real physical quantity like force and momentum. Energy is merely the calculation used by experimental physicists to quantify measurements of force and momentum. Physically, energy is the equal inseparable half of the energy/mass duality of matter $e/m = CC$.

Common Sense is a Philosophical Principle of Measurement
We sense gravity as a simple mechanical upward push. There is way too much complex, mathematical, conceptual, philosophical, and emotional baggage involved in turning the simple measured push of gravity around into a metaphysical pull that we can't even sense or measure.

We experience the world through a combination of measurements made with what we generalize as our senses. We are all aware of our five senses through which we contact and "feel" the outside world. All that we know is what we have learned from the common perception of the senses of Feeling, Hearing, Smelling, Tasting and Seeing.

Touch or feeling and balance is our most basic sense. It is the way that our bodies measure Newton's laws of force and motion $F = ma$. Our sense of feeling and balance differentiates between the directions of push and pull forces. Hearing is the way our bodies measure vibration waves passing through matter. These are sensed as repetitive push and pull forces and motions of our ear drums. Smell and taste are the passive and active components of a single sense that measures vast numbers of different individual chemical reactions that ultimately depend on interactions between electrons and protons. Our sense of sight measures individual photons from a tiny slice of the electromagnetic spectrum. They are sensed as the momentum of individual photons being absorbed and measured within our eyes.

These senses have much in common with each other but were all developed independently over the last few billion years of our genetic existence. Each of us is the final link in an unbroken genetic chain passing through each of our ancestors. The essence of what is alive within us has been part of a living organism for billions of years. While water may come and go within our bodies, there is an extremely good chance that some of the original water molecules and other atoms within the genetic structure of our bodies have been within every link of our ancestral chain from the time we were primitive creatures crawling out of the ocean. The number of links in our genetic chains is small indeed compared to the number of water and other molecules within this creature's reproductive structure. Some of these original atoms are passed at each generational link in our billion-year-old living genetic chains.

As the chain progressed from within our past, we developed our senses. Each evolved at different rates and became specialized to better suit the survival of individual organisms. As we look back at our genetic past we could say that our life began with our most distant ancestor's first sensory input. A single celled organism was first able to "see" when it could feel itself absorbing photons. Later, multi-celled organisms were able to distinguish between photons of different wavelengths. Eventually eyes were developed to measure the trajectories and energies of all the photons from a very narrow spectrum of wavelengths.

The five categories of sensory inputs are gathered together in each of us to form the common sense of our consciousness. At the center of each of our beings is the common sense of consciousness that we glean from all of our other senses. Seeing and feeling is believing. What we experience, remember, and believe in our world comes to us from combining our separate senses into the common sense of our conscious imagination. Imagination allows sensory information to be processed in non-sensory ways. Our sense of consciousness is our total reaction to the memory of our body's five sensory measurements.

Imagination can ignore information from our push and pull senses and imagine forces that are neither pushes or pulls. Our imaginations can calculate theories for forces and motions that are either beyond the range of our senses or outside of all sensory input.

We experience and interpret our sensory information in many different ways that may or may not have any relationship to the true meaning of that sensory memory. In this way, common sense and imagination are opposites. The essence of imagination is to arrange one's sensory inputs in nonsensical ways that allows us to bridge gaps in our sensory information. We see the sun moving overhead even though it is Earth that rotates under a stationary sun. It is the power of our non-sensory imagination that allowed Einstein to imagine and then believe in a form of gravitational force and motion that is the opposite of what we feel and experience with our senses. It is non-sensory imagination that allows theorists believe that photons have no mass when their eyes use the photon's individual momenta to detect and measure them. How can a photon have momentum and not have mass. Momentum without mass can be imagined but not measured. Mass is just the idea we use to quantify momentum.

Energy/mass is the structural component of matter $e/m = CC$ and photons $e/m = cC$. Energy and mass are the equal and inseparable parts of electrons, protons, and photons and neither can exist without the other. Where Einstein went wrong was when he imagined his equation $E = MC^2$ by believing he could first separate energy and mass and put them on opposite sides of the equation and then foolishly combine the rotational speed of light C with the linear speed of light c.

In experimental physics the reach of our senses has been extended many fold. Whereas our eyes can only see a narrow band of photon wavelengths with the momentum of the reddest photons only about half the momentum of violet photons, the experimental physicist can measure photons from the entire electromagnetic spectrum that have momenta and wavelengths many orders of magnitude greater and smaller than visible photons. We can measure very short gamma photons as well as extremely long photons far past the radio spectrum. With the naked eye we can barely see visible photons from galaxies like Andromeda that are only about a million light years away. Telescopes can extend our senses to a wide spectrum of photons from stars and galaxies many billions of light years away.

The enormous range of experimental measurement provides physicists with an extremely broad foundation of "common sense" on which to base physical principles that can be developed by theorists into theories. Unlike the human organism, the body of scientific physical measurement has no imagination. The common sense of physical principles does not depart from or contradict that which is measured. This principle of common sense is adhered to throughout these contemplations of physical principles. The physical parameters of measurement serve as the basis of each physical principle and the sense of imagination need only appear in a theory's final conclusions. Theories are based on imaginary metaphysical parameters that cannot be detected by physical measurement or comply to the principles of the common sense knowledge of physical evidence.

Theoretical Physicists Ridicule the Philosophy of Common Sense

The principle of common sense has been virtually ignored by Einstein and the group of people who put together the standard model theories of physics and the Big Bang. In fact, in the academic world of Harvard theoretical physics, the idea that deep physics problems might have common sense solutions is often treated with great derision. They can only imagine physics to be a deep and magical paradoxical mystery that can only be understood in terms of metaphysical assumptions and abstract mathematical equations. Only a mental midget with an insufficient imagination or a Harvard experimental physicist would attempt to explain natural phenomena just in terms of their physical measurements. In the case of Harvard's Pound-Rebka experiment, the experimental physicists made the measurements of gravitational and photon motion and then the theoretical physicists invented several different calculations to describe what was happening. Since these different counter-intuitive calculations were "equivalent" with one another, the theorists decided that they were all correct even though they contradicted the common sense values that the experimental physicists had measured.

A prime example of this kind of thinking is Einstein's famous statement, *"Imagination is more important that knowledge"*. Here he clearly shows his preference for using magical thinking and metaphysical ideas to try and contradict experimental facts. Just one example of this was his adoption of the equivalence principle so that he could imagine that the direction of gravitational force pointed downward in contrast to all experimental measurements of it pointing up. Einstein didn't even try to measure gravity. What he apparently failed to ever do was to just lay back in bed and with his sense of touch and balance and consciously feel the force of gravity pushing him upward. Because he was a daring, imaginative, and foolish fellow, Einstein ignored his own common sense experience and then went on to fool other theorists into believing in his complex and paradoxical notion that gravity extended to the farthest reaches of space instead of just to the sensors inside the seat of his pants.

Ever since Einstein was able to get away with ignoring common sense experimental principles and construct theories based of metaphysical assumptions, this method of theorizing has become the dogma of Harvard's theoretical physicists. They begin by imagining a metaphysical principle that cannot be established by experiment. They then begin constructing their theories by arranging and compromising common sense experimental facts in such a way that they are compatible with their imagined underlying metaphysics. A successful theory is one in which a large numbers of experimental facts can be arranged around a basic metaphysical principle without internal contradiction. A theory becomes suspect when contradictory conclusions or new data require the adoption of new metaphysical principles to make the old theory work. This has happened to the attraction theories of gravity that were forced by experimental measurements to imagine and adopt new metaphysical assumptions like Dark Matter and Dark Energy existing within an imagined otherwise mass-free spacetime continuum.

The principle of the gravitational expansion of mass, space and time conforms to all sensual input and establishes philosophical principles of measurement and logic that predict and confirm all measurements of gravitational phenomena. By contrast, Einstein reverses, curves and calculates the philosophical principles of measurement and logic in such a way that his imagined downward force and motion of gravity can always be thought about and calculated in precise detail but never felt or measured.

The Basic Measured Values of the Pound-Rebka Experiment

Earth radius at bottom of tower --------------------------------- $R_B = 6,371,000$ m
Earth radius at top of tower ----------------------------------- $R_T = 6,371,022.5$ m
Gravity at bottom of tower --------------------- $g_B = {}_{es}V_B^2/2R_B = 9.807$ m/s^2
Gravity at top of tower ------------------------- $g_T = {}_{es}V_T^2/2R_T = 9.8069308$ m/s^2
Acceleration of gravity at Bohr radius----------------- $g_{ao} = 8.018 \times 10^{-17}$ m/sec^2
Escape/surface velocity at Bohr radius --------- ${}_{es}V_{ao} = G_V = 9.2116 \times 10^{-14}$ m/s
Escape/surface velocity at bottom of tower--- ${}_{es}V_B = \sqrt{2gR_B} = 11,178.5864$ m/s
Escape/surface velocity at top of the tower --- ${}_{es}V_T = \sqrt{2gR_T} = 11,178.5667$ m/s
Difference in ${}_{es}V$ (${}_{es}V_B - {}_{es}V_T$) between top and bottom------------------.0197 m/s
Transverse Doppler shift of ${}_{es}V$ difference $\lambda = \sqrt{1- v^2/c^2}$ --------------2.5 x $10^{-15} \lambda$
Photon transit time between emitter and receiver 22.5m/c -----.00000007.5 sec
Velocity increase of receiver during photon transit V = gd/c--.000000736 m/s
Direct Doppler shift of transit velocity = gd/c^2 --------------------- 2.5 x $10^{-15} \lambda$
Transverse Doppler shift of ${}_{es}V_B$ $\lambda = \sqrt{1- v^2/c^2}$ --------- 1.0000000006959484 λ
Transverse Doppler shift of ${}_{es}V_T$ $\lambda = \sqrt{1- v^2/c^2}$ --------- 1.0000000006959459 λ
Difference in transverse shift between top and bottom- .0000000000000025
Clock time interval at top of tower T_T -------------------- 1.0000000000000000 s
Clock time interval a bottom of tower T_B --------------- 1.0000000000000025 s
Deceleration of gravity during photon travel time------V = 1 = .000000736 m/s

In conclusion, when the Pound-Rebka results are taken at face value, there are no experimental, philosophical or logical reasons to believe that the gravitational theories of Newton or Einstein have any validity in our own personal

Why Einstein Was an Ignorant Fool　　　　　　　　　　　*James Carter*

sensory reality. The only reason that anyone would believe in any such fanciful ideas would be the purely emotional and psychological desire to feel secure in their perception of an eternally unchanging and unmoving world.

Einstein's foolish belief in a downward pull of gravity was merely a throwback and extension to the flat Earth and geocentric theories of former times. The flat Earth people didn't want to believe Earth could curve, the geocentrists didn't want to believe Earth could move, and today, the relativists don't want to believe Earth can move from within. Even though they feel and experience it their whole lives, most foolish people are unwilling to even consider they are constantly being pushed upward to an absolutely real velocity of 11 km/s. This upward velocity and the force it produces is responsible for everything we feel and measure gravity to be. Gravity is Earth moving through the fourth dimension of space and the second dimension of time.

Centrifugal, Centripetal, and Gravitational Force and Motion

The use of accelerometers to measure the motion produced by inertial force, centrifugal force, centripetal force, and gravitational force is at the basic foundation of all experimental physics. The paradox is that these simple measurements are more or less completely misunderstood by both scientists and laypersons alike. It seems that almost no one is capable of fully understanding what these forces really are or the important role they play in our everyday realities. Everyone seems to generally imagine forces and motions as upside down and backwards and in particular, almost no one has a proper understanding of the nature and measurement of centrifugal force.

The Mathematical Dimensions of the Foundation of Reality

Space is the Non-Dimensional Negative Reality
Momentum is the One-Dimensional Reality
Angular Momentum is the Two-Dimensional Reality
Energy is the Three-Dimensional Reality
Gravitational Force and Motion is the Four-Dimensional Reality
Gravitational Time is the n-Dimensional Positive Reality

One-dimensional Forces and Motions can be on any Vector
Two-dimensional Centripetal Forces are on a Circular Plane
Three-dimensional Forces of Gravity are measured on a Spherical Plane

Mass Exists in Space = ms
Momentum Moves Mass through Space = ms/t
Space Exists as a Negative Reality and Does Note Change $s = t/m$
Time Quantifies the Motion of Mass $t = ms$
Angular Momentum Spins Mass in Space = mrs/t
Force Measures Momentum Through Time = ms/t^2
Centripetal Force Measures Angular Momentum in Time = mrs/t^2
Energy Divides Force with Velocity = $ms/t^2/2$
Photon Momentum Moves Photon Mass = mc
Photon Angular Momentum Spins Mass = $m\lambda C/2\pi$ = constant 10^{-34} kg m
Photon Kinetic Energy Moves Mass Through Space = $mc^2/2$
Photon Spin Energy Moves Mass in Space = $mC^2/2$
Gravitational Momentum Creates Orbital Motion = mrs/t
Gravitational Force Pushes Us All Up = mrs/t^2
Gravity Forces Mass Through Time and Space   Gravity = Inertia

Accelerometer Measurements of Centrifugal Force and Motion

There are 3 basic quantities in the Newtonian experimental measurement process: Mass, Space, and Time. All conceivable experimental measurements are made with Newtonian accelerometers to quantify individual values for Mass, Space, Time, and Gravity. These values are combined together in the calculations of momentum, angular momentum, force, energy and gravity. Energy is the idea used to divide a single force into multiple values. Gravity is a measure of both force and radial momentum. Whereas momentum and force exist on individual one dimensional vectors, the force and momentum of gravity are measured and calculated at the surface of a three-dimensional sphere.

Momentum and angular momentum are the fundamental measurable parameters of our physical reality. Linear energy is a calculation to give a relative quantity to momentum and rotational energy is a calculation to give an absolute quantity to angular momentum. Momentum defines absolute motion, and energy is a measure of absolute change in momentum. Momentum is a principle of measurement and energy is an assumption used to quantify both relative linear momentum and absolute angular momentum.

Accelerometers that measure force to calculate momentum, energy and velocity are the only instruments available to modern experimental physicists. All measured values eventually break down into individual changes in momentum. When we watch TV our eyes measure the individual variations in momentum of the photons emitted by the screen. When we measure gravity with accelerometers and clocks, we can only conclude that it is a three-dimensional upward pushing force that produces three-dimensional upward motion. There are no accelerometer measurements that show gravity to be a two-dimensional downward pulling force.

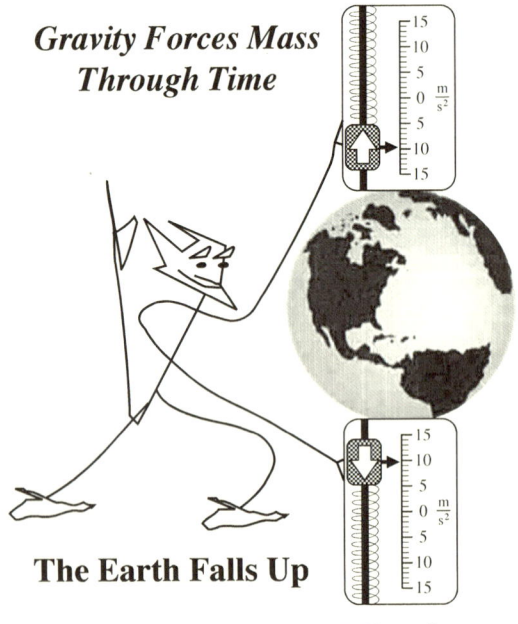

Gravity Forces Mass Through Time

The Earth Falls Up

The Upward Force of Gravity Produces Inertial Motion

Why Einstein Was an Ignorant Fool   *James Carter*

Mass, Space, Time, and Gravity

Mass, existing in Space and moving in Time is the fundamental metaphysical assumption of all physical measurements. The assumption is that Mass is eternal and when located at a position of Zero Momentum Rest, it has an absolute quantity of one and a momentum of p = 0. At this position, Mass = 1, Space = 1, Time = 1 and Gravity = 1^2. On any momentum vectors relative to rest, mass increases to M = 1+ and time intervals increase to T = 1+ proportionally to increases in momentum. Deceleration on a momentum vector decreases mass and shortens time intervals. The linear dimensions of inertial Space have a negative reality and are always calculated as remain constant. The radial dimensions of gravitational space have a positive reality are calculated to expand with gravitational force and motion.

Because they are metaphysical assumptions, mass, space, and time cannot be measured independently. The only two physical measurements that can be made of the physical interactions of mass, space, and time are relative changes in momentum and absolute changes in angular momentum.

Every moving body has an unknown absolute momentum vector that is a combined unit of mass, space, and time. When we change and measure a body's momentum, it is done with a force that changes its absolute momentum to a new unknown vector and value. This new measured momentum vector is produced by an unknown combination of acceleration and deceleration. When we change and measure a body's angular momentum, it is done with a centrifugal force that either produces acceleration or deceleration. Changes in momentum are measured with accelerometers on a single vector as Force F = ma*d^2. Force equals mass times an unknown combination of acceleration and deceleration.

E = MC² is Wrong and E/M = cC & E/M = CC are Right

Energy and mass are two sides of the same coin and are always equal and cannot be physically separated in any conceptual way. e/m = cC is the formula for photons and e/m = CC is the formula for atoms. A moving body's value for energy/mass is e/m = v^2 - c^2/c. At rest, energy/mass equals zero, e/m = 0^2 - c^2/c = c/0. At the speed of light, energy/mass would become infinite e/m = c^2 - c^2/c = 0/c.

The force produced by an energy always produces two equal and opposite quantities of momentum p = mv = MV and two unequal quantities of energy e = mv^2/2 and e = mV^2/2. Kinetic energy is quantified by dividing e = mv^2 into two unequal parts. The total energy of a force is measured with two accelerometers that divide the force into two equal parts and energy into two unequal parts.

Energy/mass, is a conserved eternal quantity. The quantity of mass in the universe is always equal to its total energy e/m = c² and neither quantity ever changes separately. This is because they are two parts of a whole and cannot be separated from one another in any meaningful way. When a body is accelerated to increase its energy, its mass is also increased by an equal amount. Mass is energy and energy has mass.

All photons have a momentum of p = mc and an angular momentum of h/2π = mλC/2π = mCr. A photon is a matter/antimatter duality composed of a length positive magnetic cosmic string from a proton combined in a wavelike motion with an equal length of negative electric cosmic string from an electron. The angular momentum of a photon's back and forth wave motion is a universal constant of nature. The two pieces of cosmic string are also spinning at C in opposite directions, with equal quantities of angular momentum h/2π = mCr. When these two opposite spins are added together, the total angular momentum from spinning strings is zero. However, total energy of a photon's spinning strings e = mC²/2 can be measured and combined with the energy of its linear motion e = mc²/2 for a total photon energy of e = mc²/2 + mC²/2 = mcC. The photon's momentum p = mc provides 1/2 of its measured energy and the angular momentum of its rotating mass provides the other half e = Iω²/2. The energy of its momentum is relative to an observer's motion and the energy of its angular momentum remains constant for all observers.

Momentum produced by a force can only be measured relative to the rest frame of an accelerometer and total energy produced by a force can only be measured relative to the rest frames of two accelerometers. The energy inherent in angular momentum e = mωr²/2 is an absolute and constant quantity that is measured to have the same value in the moving positions of any observers. When the values of mass, rotational velocity, and radius are changed to maintain constant angular momentum, the rotational energy is also changed. As with momentum, angular momentum contains a body's kinetic energy but it is not a measure of it. Different spinning bodies can have the same values of angular momentum but greatly different rotational energies.

The Absolute Accelerometer Measurements

Accelerometers measure the force producing the acceleration or deceleration of mass. They are the only possible measuring devices for quantifying changes in a body's momentum or angular momentum. Modern experimental physicists have thousands of different measuring devices for determining all of the many different quantities in physics. However, at their most fundamental level, each of these instruments measures and calculates changes in momentum and angular momentum through the measurement of accelerations and decelerations produced by force. As an example, consider the radar gun used by police to measure the speed of cars. This device determines traffic speeds by measuring changes in the momentum of photons reflected from moving vehicles. The Hubble red shifts are measured as changes in the momentum of spectral photons. All of an experimental physicist's many measured and calculated values are based on accelerometer readings of either a linear force or centrifugal force that changes momentum or angular momentum.

Momentum, Force and Energy of Cannons and Cannonballs

In the following thought experiment with a cannon and golden cannon ball, their individual momentum is easy to calculate because they are always equal. However, the recoil energy from the force of the gunpowder is much greater for the ball than the cannon. Momentum is measured from the acceleration produced by of a single quantity of force and energy is calculated from the opposite accelerations produced by force. In this sense, momentum is absolute and energy is relative even though they are both always equal. Momentum is measured as motion in a single absolute frame and energy divides the motion between two relative frames.

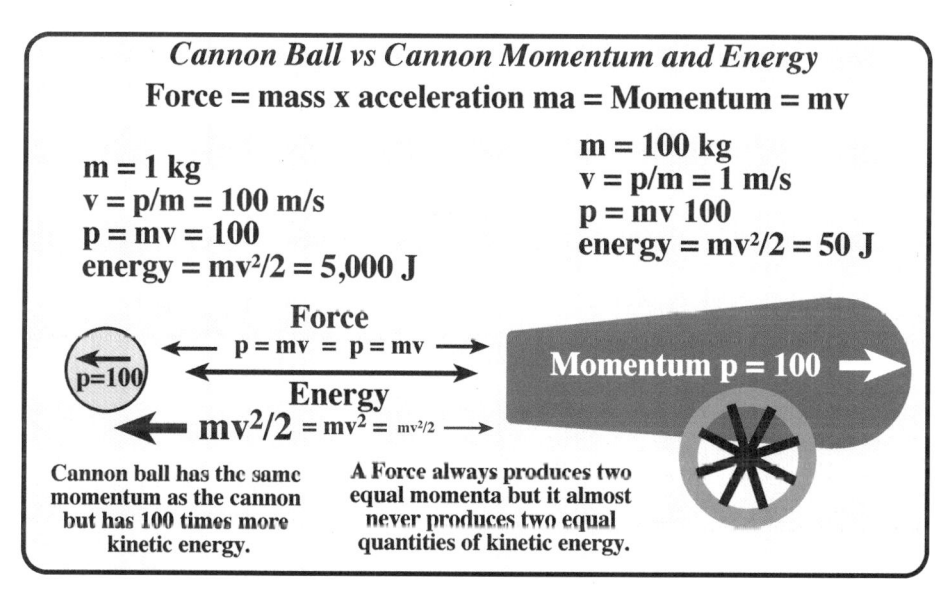

Cannon Ball vs Cannon Momentum and Energy
Force = mass x acceleration ma = Momentum = mv

$m = 1$ kg
$v = p/m = 100$ m/s
$p = mv = 100$
energy $= mv^2/2 = 5{,}000$ J

$m = 100$ kg
$v = p/m = 1$ m/s
$p = mv$ 100
energy $= mv^2/2 = 50$ J

Force
$p = mv = p = mv$
Energy
$mv^2/2 = mv^2 = mv^2/2$

p=100

Momentum $p = 100$ →

Cannon ball has the same momentum as the cannon but has 100 times more kinetic energy.

A Force always produces two equal momenta but it almost never produces two equal quantities of kinetic energy.

Since we are unable to measure the cannonball's momentum without changing it, we can calculate its exact velocity and energy by measuring the cannon's equal momentum. If the cannon is rifled, and causes the cannonball to spin, then calculating the total energy of the ball, when it hits the target, becomes more complicated. The cannonball's linear energy $e = mv^2/2$ is strictly relative to its motion with the target. However, its rotational energy $e = m\omega r^2/2$ is absolute and has the same value at all moving targets.

Measuring of Positive and Negative Centrifugal Forces
Centrifugal Force is a Momentary Transverse Push.
Centrifugal forces are measured as momentary transverse accelerations or decelerations. Whereas, a linear force that changes momentum is on a vector relative to the accelerometer's position of rest, centrifugal forces that change angular momentum are perpendicular measurements relative to the center of a two-dimensional plane.

Centripetal Force is a Constant Two-Dimensional Radial Pull.
Centripetal force is a constant and equal balance of radial acceleration and deceleration. Centripetal force can be measured at anyplace on the table with two accelerometers set at 90° apart and aligned with the center. One measures a constant one-half transverse acceleration and the other measures the equal one-half of a transverse deceleration.

Both centrifugal force and centripetal force are measured with accelerometers attached to a rotating body. Centripetal force is a measure of the total positive centrifugal forces that have created the rotation. A free-wheeling rotating turntable produces a constant centripetal force at any location on the table. Transverse centrifugal forces produce either positive or negative changes in the turntable's angular momentum and rotational energy. Momentary centrifugal forces are equally added to or subtracted from the table's constant centripetal force.

The turntable measurements of force and momentum are done with two accelerometers. Centrifugal force accelerometers are aligned at 180° and centripetal force accelerometers are aligned at 90°. Centrifugal forces are measured as positive and negative one-dimensional pushes that are equal and opposite to the two-dimensional centripetal pulling force. These two forces are really just the same force aligned at 90° from one another. The momentum and linear energy produced by centrifugal force are stored in the angular momentum and rotational energy of centripetal force.

Change in Momentum Produces Force that is Measured as Energy
It is a common misconception of both scientists and laymen alike when they say, "To increase the rotation of the turntable we must add energy to it." This statement is true only in a relative sense. When we measure and change the rate of the turntable's rotation it is done by adding equal, consecutive units of centrifugal force. Each unit of force increases or decreases the table's centripetal force, angular momentum, and rotational energy. Force is the cause of momentum change and energy is its measurement. It should be clear from this, that when we accelerate the table by using centrifugal force to equally increase

its angular momentum and centripetal force, we are adding kinetic energy in the table.

This rotating turntable experiment begins at Position 1 with the balls resting in a radial groove. Ball #1 is just leaving the table at v = 4 and the table is spinning at v = 4. The total angular momentum and energy of the table and the balls is $I\omega = 1$. The table's centripetal force is $F = 1$ and its centrifugal force is $F = 0$.

When the balls are all released from centripetal pulling force, centrifugal pushing forces becomes active and begin pushing the balls out the groove. As the balls move out toward the edge of the table, their velocity and energy are increased by the centrifugal push from the back edge of the groove in the table. Accelerometers on the balls measure a positive centrifugal pushing force from the turntable and accelerometers mounted to the table measure an equal and opposite negative centrifugal pushing force that decreases table's rotational

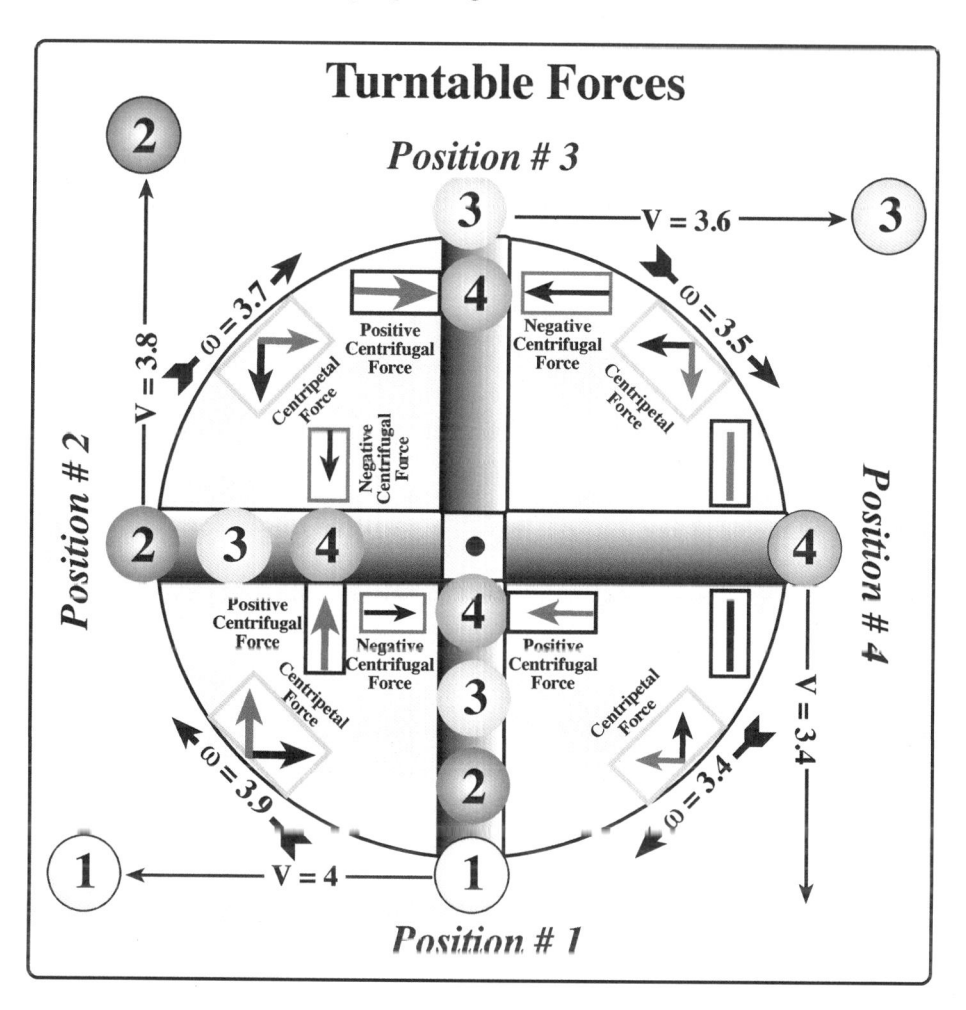

Why Einstein Was an Ignorant Fool   *James Carter*

velocity. The centrifugal force converts rotational energy from the table to the linear energy of the accelerating ball. In this process, the table is slowed but its energy and angular momentum remain constant until the next ball is released.

As the balls roll out to Position #2, angular momentum is being transferred to the accelerating balls from the decelerating table. As the balls' velocity speeds up, their energy increases, and as the turntable's velocity slows its energy decreases by an equal amount. Total combined angular momentum and energy of the table and the balls remain constant.

In this experiment, accelerometers are used to measure the positive and negative centrifugal forces that change the angular momentum and energy of both the table and the balls. The is a passive measurement that involves no change in total angular momentum or kinetic energy. What is measured are the centrifugal forces required to transfer angular momentum and energy from the table to the balls.

This experiment consists of measurements at eight positions on a single rotation of the table and balls. Four positions measure the table's decreasing centripetal force as the balls move along the grooves and the other four positions measure proportional decreases in positive and negative centrifugal forces. The arrows represent accelerometer readings with their length indicating their equal force and their size indicating the amount of acceleration or deceleration produced. Red arrows represent accelerations and black arrows are decelerations. All arrows are measurements of the forces that change the balance of angular momentum between the table and the balls. As the Balls leave their grooves, their angular momentum is converted into linear energy of momentum. However, even after the balls leave the groove in the table and are far away, the total angular momentum and energy of the balls and table remains conserved.

Lorentz Transformations of Mass to Energy

At the end of the 19th century, Hendrik Lorentz predicted that a body's inertial mass would increase as it was accelerated to higher and higher velocities. This increase in kinetic mass was proportional to the formula $\sqrt{1-v^2/c^2}$, while the relationship between a moving body's kinetic energy $e = mv^2/2$ and its increased kinetic mass was $m_K = e/c^2$.

What this transformation means is that when we accelerate an automobile to a high velocity, both the car and our body increase in mass in proportion to the kinetic energy from speeding down the highway. In this case, the change in mass is very small, but by comparison, the mass inherent in our kinetic energy begins to become significant when we consider the absolute velocity of 375 km/s that our bodies have relative to the zero momentum rest frame identified by the 2.7°K Cosmic Blackbody Radiation dipole anisotropy. This velocity gives a 100 kg person a kinetic energy of 1.8×10^{13} Joules and a kinetic mass increase of about 200 milligrams. This absolute velocity relative to zero momentum rest gives us each real linear kinetic energies that are about equal to the first atomic bomb that was exploded in New Mexico (10^{13} Joules) (see *Joules of the Universe*). This energy is completely hidden from us and could only be realized by the negative kinetic energy needed to slow us down to a stop. While this velocity might at first seem to be very fast, it is only about 1/1000th the speed of light. For our bodies to double in mass, we would have to travel in a spaceship at 87% the speed of light. At this velocity, our rest mass and kinetic mass would be equal and we would each have personal kinetic energies equal to about one million atomic bombs (10^{19} Joules).

The Lorentz transformation experiments listed below show that for every change in a body's velocity and energy there is complementary change in its mass. In an atomic explosion, about one gram of matter seems to change into the kinetic energy (10^{11} J) of moving bodies of mass and photons. The daughter atoms left over from fission have less mass than the original uranium atoms mostly due to the loss and subsequent decay of neutrons. The energy produced in the explosion consists of photons and decaying neutrons, as well as the kinetic energy of rapidly moving particles and atoms. Nowhere in this transformation is mass converted into energy. Energy/Mass are constant and never change $e/m = c^2$.

The energy used to accelerate all of the particles in the explosion (heat) has created a Lorentz transformation in mass, meaning that their mass and energy remain constant before and after the explosion. In the same way, the mass of the matter converted into the energy of photons decreases by the same amount as the photons' mass $m = e/c^2$. This transformation of rest mass into the kinetic mass of energy is as common as lighting a match or kicking a football. It is just that in such cases, the energy is easy to measure but the mass change is too small to be detected. Mass and energy are two sides of the same coin and are always present in equal quantities and never separate.

The Idea of Lorentz Length "Contraction"

The three parameters of mass, space and time are all involved in the Lorentz transformation measurements of kinetic mass, time dilation, and the idea of length contraction.

Einstein adopted the Lorentz transformation as a principle of physical measurement in both the special and general theories of relativity. It is not an actual part of his theories because it is a physical principle and not an unmeasured metaphysical assumption like his theories of the massless photon and the equivalence of inertial and gravitational motion.

Of the Lorentz transformation predictions, mass increase and time dilation have been accurately measured but the predicted length contracted caused by increased velocity has never been detected.

The following thought experiment demonstrates why length contraction is not real and how the idea arises from the experimenter's choice of constants and physical parameters in the measurement process.

In the experiment, a 300,000 km ruler with a mirror at the end is attached to a space ship. Photons are emitted and then reflected back to the ship by the mirror. Before the ship takes off, it is measured that it takes two seconds for the photons to return to the ship. However, when the ship and ruler are moving at 87%c it only take one second for the photon to travel to the mirror and back. This is because the observer's clock has slowed to one half its Earth rate. If the observers don't realize the clock has slowed and assume the speed of light to be constant, they will incorrectly assume that the ruler has contracted to one half its Earth length.

The slowing of the clock is caused by the doubling of mass in its rotating parts without any change in their angular momentum. Believing in length contraction comes from not realizing the clock's mass and intervals have increased.

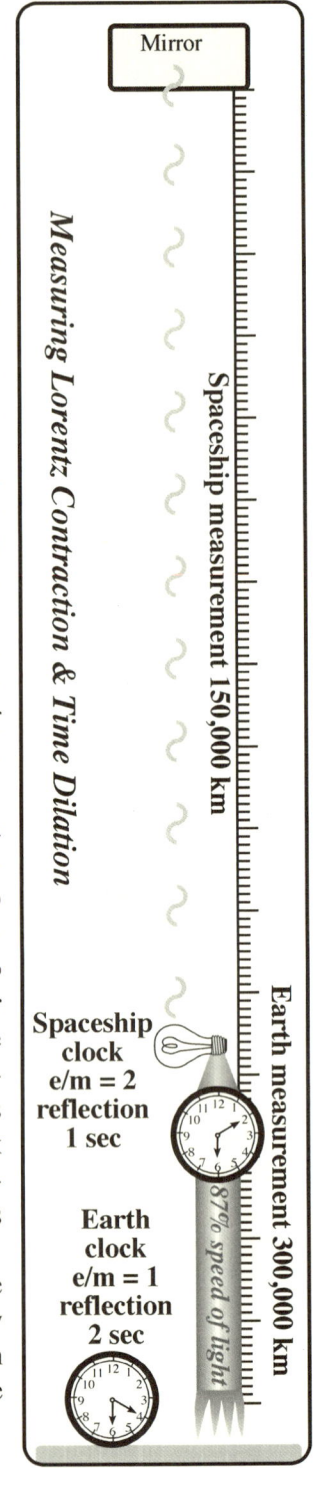

Measuring Lorentz Contraction & Time Dilation

Spaceship measurement 150,000 km

Earth measurement 300,000 km

87% speed of light

Spaceship clock
e/m = 2
reflection
1 sec

Earth clock
e/m = 1
reflection
2 sec

Why Einstein Was an Ignorant Fool   *James Carter*

Experimental Values of Lorentz Transformations of Mass and Time

Lorentz Transformation Thought Experiment
GPS Clock Calculations
Pound-Rebka Experiment
Triplet Paradox Experiment

The Lorentz transformation m' = M/√1-v²/c² is a principle of measurement that can be classed as one of the laws of physics. (A moving body's kinetic mass m' is equal to its rest mass M divided by the square root of one minus the Lorentz velocity squared v² divided by the speed of light squared c²). It comes into play whenever a body of mass undergoes measurable acceleration or deceleration. This equation calculates the changes in a body's mass that occur with measured changes in its momentum.

This equation is simply a different version of E = MC². The photon version of this equation is C² = E/M where energy and mass are always equal. As the momentum of a body is increased, its energy and mass are increased by equal amounts. As the body approaches the speed of light the measured value of its energy/mass increases requiring more and more kinetic energy to accelerate it faster. At the speed of light the body's energy/mass would become infinite. The Lorentz transformation for the energy/mass of the photon is m' = M/√1-c²/c² = E/M.

The inverse of this equation t' = T /√1-v²/c² (T² = E/M) calculates the length of a clock's time intervals as it slows down or speeds up in direct proportion to increases or decreases of the energy/mass of the clock's momentum vector relative to the Zero Momentum Frame of rest. When a clock is accelerated, its mass increases and its time intervals grow longer at the same rate. When a clock is decelerated, its energy/mass decreases to a minimum of m = 1 at rest and its time intervals grow shorter to a maximum rate of t' = 1.0 where the clock is at absolute rest and has no momentum to dilate its intervals.

This change in the duration of clock intervals is simply an effect of mass changes in the clock's internal mechanism. As a clock's mass (momentum) increases with increasing velocity, the angular momentum of its rotating and vibrating components remains constant. It is this conservation of angular momentum that causes these components to slow their motion and increase the intervals of time that they record. This process is reversed when the clock is decelerated and the lengths of its recorded intervals are decreased.

The Lorentz transformation is a principle of measurement and not a theory. It is the calculation that we use to determine the changing values of mass and clock time intervals whenever acceleration and deceleration are measured. While Einstein incorporated the Lorentz transformation into both of his relativity theories as a principle of measurement, it was not a structural part of the metaphysical assumptions made in either the special of general theories of relativity. The Lorentz transformation is based on Newtonian accelerometer measurements that Einstein used to interpret his metaphysical assumptions about the force and motion of both photons and gravity.

Both m' = M/$\sqrt{1-v^2/c^2}$ and t' = T/$\sqrt{1-v^2/c^2}$ determine the inertial frame for each body of mass such as a clock. There are an infinite number of inertial frames with different values for their Lorentz momentum velocity vector p = mv, but the universe contains only a single zero momentum Lorentz frame where p = 0, v = 0, m' = 1, M = 1, t' = 1 and T = 1. All clocks with the same Lorentz velocity (v) have the same mass increase and increased clock intervals regardless of the direction of their motion. It is the absolute velocity of a clock's momentum vector that determines its mass value and time intervals and not the relative velocity between any two bodies. Two clocks can be moving side by side at v = x and have no relative velocity between them or they can be moving in opposite directions with a relative velocity of v = 2x. In both cases the values for their mass and time intervals will be the same.

Even though the relative motion between bodies in two moving frames has no effect on their mass and time, it is the only component of each body's Lorentz velocity vector that can be measured. The experimental process is unable to separate measured relative acceleration into its separate components of absolute acceleration and deceleration that produce changes in a clock's mass and time intervals.

Lorentz Transformation Thought Experiment

Imagine two pairs of spacecraft containing Cesuim-133 clocks and technicians. Each pair of craft is separated by some distance and moving toward one another at a relative velocity of 1km/s. One space craft is at zero momentum rest with a velocity of v = 0 while the other is moving toward it at v = 1 km/s. The second pair of craft are moving nearly side by side at v = 150,000 km/s (1/2 c) and v = 150,001 km/s respectively. From their relative motion, we must conclude that each pair of clocks is moving along separate momentum vectors that are nearly identical. In each case, the technicians measure them to be moving toward one another with an average relative velocity of v = 1 km/s.

In the course of the experiment, the two crafts move closer together, pass, and then move farther apart. The purpose of the experiment is to acquire information about the true absolute motion of each clock. The technicians use Doppler shifted photons to monitor their changing relative motion as they pass. This relative motion measured with Doppler shifts is only valid for individual points in time. Photons are blue-shifted as the clocks approach and then are red shifted as they recede from one another. At the time interval when the two ships pass, there are no Doppler shifts between them (except for transverse shifts) indicating they have no relative motion. However, when all of the Doppler measurements are calculated together, it is determined that the two ships's average relative velocity is v = 1km/s.

At the point where the spaceships pass, it is easy for the technicians to compare the difference in their clocks' intervals and determine a portion of their true momentum vector. If one clock is actually at rest with a mass and time interval of 1.0 and the other has an absolute velocity of v = 1km/s, then the mass and time interval of the moving clock would be m' & t' = 1.0000000000056. However, with one clock moving at 150,000 km/s and the other moving at 150,001 km/s, then the first clock would record time intervals of 1.15470054 and the second clock's intervals would be 1.15470310. The difference in clock rates between a v = 1km/s relative velocity at rest and a v = 1 km/s relative velocity at ½ c is enormous. The clock interval increase for 1km/s at rest is more than 5 orders of magnitude smaller than the difference in intervals for v = 1km/s of relative motion between clocks moving at v = 1/2 c.

Lorentz Transformation Mass and Time Values for 1/2 c

Mass of Clock at 1 km/s--------------- m' = $M/\sqrt{1-v^2/c^2}$ = 1.0000000000056 kg
Clock interval for 1km/s ------- t' = $T/\sqrt{1-(1 \text{ km/s})^2/c^2}$ = 1.0000000000056
Mass of Clock at 150,000 km/s------ m' = $M/\sqrt{1-v^2/c^2}$ = 1.15470054 kg
Mass of Clock at 150,001 km/s------ m' = $M/\sqrt{1-v^2/c^2}$ = 1.15470310 kg
Clock interval for 150,001 km/s -t' = $T/\sqrt{1-(150,001 \text{ km/s})^2/c^2}$ = 1.15470310
Clock interval for 150,000 km/s -t' = $T/\sqrt{1-(150,000 \text{ km/s})^2/c^2}$ = 1.15470054
Difference in clock intervals of 150,001 km/s & 150,000 km/s ----.00000256
Difference in clock intervals of v = 0 km/s & v = 1 km/s-----.0000000000056
Difference in mass increase for v=1 between v =1 and v =150,001-- 457,142

The Zero Velocity Lorentz Transformation Frame

The idea of a zero velocity Lorentz frame t' = $T /\sqrt{1- 0^2/c^2}$ is a metaphysical principle for the mass and clock intervals of m', M, t', & T, all = 1.0. This zero velocity metaphysical frame is just a featureless void of empty three-dimensional space that can never be measured because it has no physical parameters. There are an infinite possible number of other Lorentz transformation frames that can be measured with clocks and accelerometers. Each frame has a different value for its momentum vector (p = mv) and a different time interval (t'/T) for its clock. These frames all share relative motion with the single zero velocity Lorentz clock frame. In all moving frames, mass and time intervals have equal Lorentz values of (m' = 1+) & (t' = 1+). Increasing the velocity of a clock increases its mass and momentum and the conservation of angular momentum in turn increases the length of its time intervals.

Why Einstein Was an Ignorant Fool   *James Carter*

Two actual experiments that use the Lorentz transformation principle to calculate the mass and time differences between two Lorentz transformation frames are the GPS clock measurements and the Pound-Rebka measurements of gamma photon momentum.

Global Positioning System Clock Calculations

In the GPS measurements, clock intervals between two different Lorentz velocity frames are calculated and measured. The first frame is the combined velocity vector of the rotational (orbital) velocity ($_o v$ = 448 m/s equator) at Earth's surface and the perpendicular upward gravitational escape/surface velocity of Earth's surface esv = $\sqrt{2gR_E}$ = 11,189 m/s. The Lorentz transformation velocity $v = \sqrt{_{es}v^2 + _o v^2}$ at Earth's equator is 11,198 m/s. The second frame is the combined vector of the GPS satellite's orbital velocity $_o V$ = 3868 m/s and the vertical upward gravitational escape velocity $_{es}V$ = 5471 m/s at its orbit. The Lorentz transformation velocity $v = \sqrt{_{es}V^2 + _o V^2}$ of the 24 GPS satellites is 6700 m/s. The relative velocity between the ground and the satellite is 11,198 − 5471 = 4498 m/s. While the relative velocity can be measured with photon Doppler shifts, this velocity is not used to calculate satellite clock adjustments. The measured value for clock adjustment is obtained by subtracting the calculated interval of the ground clocks momentum vector from the Lorentz velocity interval of the GPS clock.

Earth Clock and GPS Clock Experimental Values

Mass of 1kg clock at relative motion- $m' = M/\sqrt{1-v^2/c^2}$ = 1.000000000113 kg
Relative velocity interval----------------- $t' = T /\sqrt{1-v^2/c^2}$ = 1.000000000113
Mass of Earth clock at 11.2 km/s--------- $m' = M/\sqrt{1-^2/c^2}$ = 1.000000000697 kg
Mass of 1 kg GPS clock at 6.7 km/s--- $m' = M/\sqrt{1-v^2/c^2}$ = 1.000000000249 kg
Earth Clock's velocity interval ----$t' = T /\sqrt{1- (11.2)^2/c^2}$ = 1.000000000697
GPS Clock's velocity interval ------ $t' = T /\sqrt{1-(6.7)^2/c^2}$ = 1.000000000249
Interval slowing needed to synchronize GPS clocks -------- .000000000448

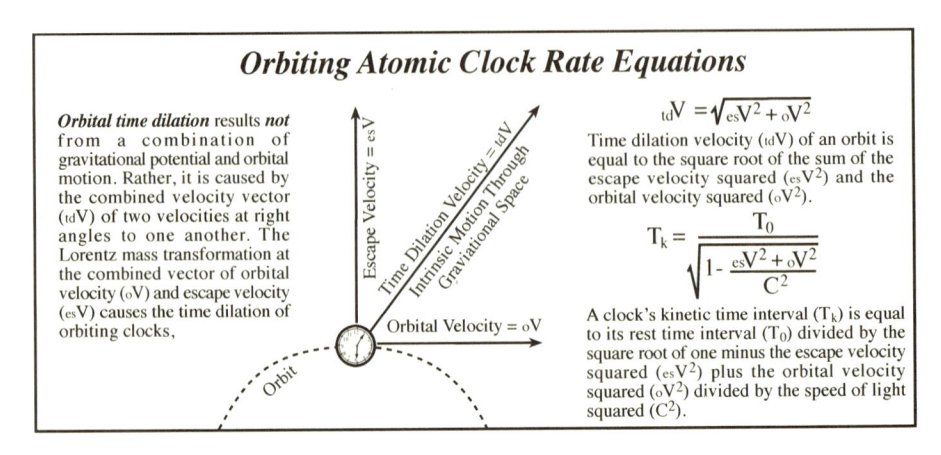

Orbiting Atomic Clock Rate Equations

*Orbital time dilation results **not** from a combination of gravitational potential and orbital motion. Rather, it is caused by the combined velocity vector ($_{td}V$) of two velocities at right angles to one another. The Lorentz mass transformation at the combined vector of orbital velocity ($_oV$) and escape velocity ($_{es}V$) causes the time dilation of orbiting clocks,*

$_{td}V = \sqrt{_{es}V^2 + _o V^2}$

Time dilation velocity ($_{td}V$) of an orbit is equal to the square root of the sum of the escape velocity squared ($_{es}V^2$) and the orbital velocity squared ($_o V^2$).

$$T_k = \frac{T_0}{\sqrt{1- \frac{_{es}V^2 + _o V^2}{C^2}}}$$

A clock's kinetic time interval (T_k) is equal to its rest time interval (T_0) divided by the square root of one minus the escape velocity squared ($_{es}V^2$) plus the orbital velocity squared ($_o V^2$) divided by the speed of light squared (C^2).

Why Einstein Was an Ignorant Fool — *James Carter*

These different clock rates have nothing to do with relative motion. For example, the 24 satellites in the GPS constellation are all moving at many different relative velocities yet their clocks all remain synchronized because their frames all have the same measured Lorentz velocity. These calculations will not be correct if the zero velocity Lorentz rest frame is used for any of these frames. It is only used as a reference for a T = 1.0 clock interval at Earth's center.

These calculations are not made with the equations of General Relativity theory. General Relativity's calculations are based on the metaphysical assumption of undetectable gravitational potentials. These potentials are derived from measured gravitational accelerations and escape/surface velocities but they cannot be measured independently. It matters not whether you use calculated gravitational potentials or measured escape/surface velocities in your calculations. The results will come out the same either way because the calculated potentials are derived from measured velocities and accelerations. The metaphysical assumption of imaginary gravitational field potentials is not needed to calculate the correct GPS clock rates.

Pound-Rebka Calculations

In the Pound-Rebka measurements, the momentum and clock time intervals at the top of the 22.5 m high Jefferson tower are compared with the greater momentum and longer time intervals at the bottom of the tower. The gravitational escape/surface velocity at the top tower is $_{es}V$ = .01974 m/s less than the escape/surface velocity at the bottom. The Lorentz transformation between top and bottom escape/surface velocities is used to measure the decreased time intervals of top clock and the increased wavelengths of absorbed photons. Einstein called the difference in clock intervals and photon wavelengths between the top and bottom gravitational red shifts.

Pound-Rebka Experimental Values

Gravitational velocity at the top of tower ------------------- 11,189 m/s
Gravitational velocity at the bottom of tower --------------11,189.01974 m/s
Difference in velocity during photon travel time-----------.000000736 m/s
Mass of clock for relative velocity – 1.0000000000000000000000000000003
Relative velocity interval---------- t'=1.0000000000000000000000000000003
Mass of top clock ------------------ m' = $M/\sqrt{1-v^2/c^2}$ = 1.0000000006959459 kg
Mass of bottom clock --------------m' = $M/\sqrt{1-v^2/c^2}$ = 1.0000000006959484 kg
Bottom Lorentz velocity interval-- t' = $T/\sqrt{1-v^2/c^2}$ = 1.0000000006959484
Top Lorentz velocity interval ------ t' = $T/\sqrt{1-v^2/c^2}$ = 1.0000000006959459
Difference in time intervals between the top and bottom--.0000000000000025
Pound-Rebka measured momentum, wavelength & interval shifts---2.5 x 10^{-15}

Why Einstein Was an Ignorant Fool   *James Carter*

If we use a zero velocity Lorentz frame to calculate the momentum and time shifts in the gamma photons and clocks used in the experiment, we get a result that is 17 orders of magnitude smaller than the measured effect. In order to duplicate the measured values of 2.5×10^{-15} for photon momentum and time dilation, we must calculate the time dilation of the two escape/surface velocity frames at the tower's top and bottom. The relative velocity of 7.36×10^{-7} m/s can be used to calculate the Doppler shifts but the actual cause of the shifts is the difference on gravitational momentum and clock intervals between top and bottom.

Again, these calculations are not based on either Special Relativity or General Relativity. These results are derived completely from the measured parameters of gravitational force and motion. They have nothing to do with metaphysical assumptions about the relative motion between the undetectable potentials of gravitational fields. *(there is a more detailed explanation of the Pound-Rebka experiment elsewhere in this book.)*

The Triplet Paradox Experiment
There is a third type of Lorentz transformation experimental measurement that makes a comparison between not two but four or more Lorentz velocity frames.

One example of this is the so called Twin Paradox experiment where one twin stays home in an assumed zero velocity Lorentz frame and the other goes on a long, high velocity, journey into space and back. When the astronaut twin returns home, he is younger than his brother due to the difference between the unchanging clock intervals of Earth's Lorentz rest frame and the increased length in clock intervals of the outbound or inbound Lorentz velocities. If both legs of the journey are at the same measured velocity, then the dilated clock time intervals will be the same for the back and forth portions of the trip.

The glaring problem with calculating the results of a twin paradox experiment is that the actual zero velocity Lorentz frame cannot be easily located and Earth's true Lorentz velocity frame cannot be located beyond comparing Earth's location with the motion of bodies in the universe in general and the Doppler shifts of 2.7°K Cosmic Blackbody Radiation photons in particular.

The thought experiment illustrated is performed by identical triplets. Adam, Bob & Chad. Adam stays on Earth for two years and watches his clock. Bob uses his accelerometer to measure an acceleration to 375 km/s in the direction of the constellation Leo, maintains that velocity for one year and then records the time on his clock, turns around and accelerates to 750 km/s back towards Earth. This gives him a relative velocity with Earth of V = 375 km/s.

Chad accelerates to 375 km/s in the direction of Aquarius and then after one year he records the elapsed time on his clock, turns around and accelerates to 750 km/s back towards Earth. This also gives him a relative velocity with

Earth of V = 375 km/s. Both triplets spend two years traveling at a velocity of V = 375 km/s relative to Earth.

If we use relativity's time dilation formula t' = t/√1-v²/c² to calculate the clock rates for 375 km/s, we find the intervals of Bob's and Chad's clocks are t' = 1.000000781 versus Adam's zero velocity intervals of t' = 1.0. These measured values are only valid for the special situation where Earth is at rest in the zero momentum Lorentz frame. Common sense tells us that Earth cannot possibly be in the zero momentum frame. If nothing else, we can see Earth moving relative to the sun. Earth's true Lorentz velocity must remain unknown until the triplet paradox experiment has been completed. By measuring the difference in intervals between each leg of an astronaut's journey, it is possible to measure the magnitude of Earth's velocity along the vector of the twin's journey. Only if the two intervals are the same can we determine that Earth is at rest along that vector.

Now, if we use the formula to calculate Earth's values within a 375 km/s Lorentz frame with its vector between Leo and Aquarius, we get different intervals for all three clocks. Adam's Earth clock with an assumed interval of T = 1.0 is now calculated to have a Lorentz interval of t' = 1.000000781.

On Bob's trip toward Leo, he is actually traveling at 750 km/sec relative to the zero velocity Lorentz rest frame. This increases the momentum of his atomic clock and increases its interval to t' = 1.000003125. This is four times slower than Adam's slowed Earth clock. After one year, Bob turns around, marks his clock, and accelerates to 750 km/s toward Earth to obtain a relative velocity of 375 km/s. In actuality, this is all deceleration that brings Bob to the position of zero momentum rest. On Bob's "trip" back to Earth, he is actually sitting still while it is Earth that is traveling at 375km/s to meet him. With no momentum to slow his clock, Bob's clock is running faster than Adam's clock with a time interval of t' = T /√1–0²/c² = 1.0. Bob can determine he is actually at v = 0 Lorentz rest by observing no dipole anisotropy in the temperature and momentum of the 2.7°K CBR photons.

When Chad accelerates to 375 km/s in the opposite direction towards Aquarius, he is actually decelerating to a stop. This decreases his clock's momentum to zero and causes it to run at its maximum rate of t' = T /√1–0²/c² = 1.0. He also will not be able to measure any dipole anisotropy in the momenta of 2.7°K CBR photons. He sits at rest for one year while Earth moves away from him at v = 375 km/s. He then marks the time on his clock and accelerates to a momentum vector of v = 750 km/s towards Earth, giving him a relative velocity of V = 375 km/s. His clock will now have an interval of t' = 1.000003125 for his true Lorentz velocity of v = 750 km/s.

In these calculations, Bob and Chad spent half their journeys sitting at rest with clock intervals of t' = 1.0 and the other half moving at v = 750 km/s with clock intervals of t' = 1.000003125. The average momentum frame time dilation interval of the traveling triplets is t' = 1.000001562. Relative to Adam's unmeasured Earth clock's rate of t' = 1.000000781, the traveling triplets will measure their average time dilations for both their trips to be t' = 1.000000781. By recording the elapsed time on their clocks when they turn around, Bob and Chad will be able to determine the correct clock time intervals for each leg of their journeys. This will allow the triplets to determine that the true motion of Earth is 375 km/s towards Leo without using the CBR dipole anisotropy as a reference.

While it is true that special relativity's twin paradox calculations for a zero velocity frame and the calculations for an arbitrarily moving Earth rest frame yield identical results for the total time dilation of Bob's and Chad's round trip journeys, they give greatly different results for the time intervals of individual legs of the triplet's journeys. When the triplets record the time intervals for each leg of their journeys, they find that they are not equal even though they were careful to maintain a precise relative velocity of V = 375 km/s with Earth. This relative velocity can be measured and verified with both inertial navigation accelerometers and photon Doppler shifts between Earth and the spacecraft. The triplets can then use the measured differences in these two clock intervals to calculate the true value for Earth's velocity along their vector.

These results clearly show that the Living-Universe's zero momentum Lorentz frame is the only possible Preferred Frame and all of special relativity's arbitrary "rest" frames can be measured to be moving frames with unique momentum vectors. While special relativity theory is able to calculate the correct average time interval for two-way trips, it fails completely to give correct time intervals for individual one-way legs of the trips without incorporating Earth's true Lorentz momentum vector p = mv.

The Zero Momentum Preferred Rest Frame

The measured time interval of 1.000000781 second, hour, year etc. tells us that Earth is moving at 375 km/s along the same x vector of the triplet's trips. The difference in time intervals between each leg of the triplet's trips determines the direction and magnitude of Earth's Lorentz velocity. There are three vectors of motion and this measurement only identifies Earth's momentum vector along its single x vector. Earth's true Lorentz xyx vector velocity must be at least v = 375 km/s but could be either much more or less along its y & z vectors. Earth could be moving up or down or sideways left or right faster or slower than it is moving along its x momentum vector.

In a triplet paradox experiment where Adam, Bob and Chad travel along the x, y, and z vectors, it would be possible to determine Earth's true Lorentz momentum vector relative to the Zero Momentum Frame m' = M/$\sqrt{1-0^2/c^2}$ = 1 kg. The ZMF is the measured preferred absolute rest frame of the Living-Uni-

Triplet's Journey in 2.7°K Cosmic Blackbody Radiation Time

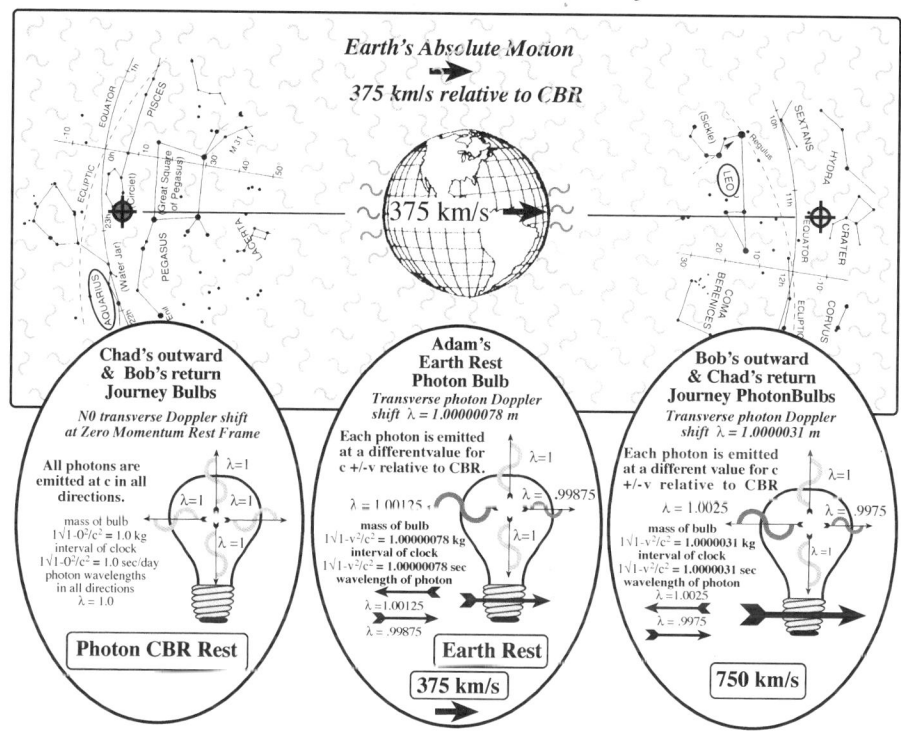

Earth's Absolute Motion
375 km/s relative to CBR

375 km/s

Chad's outward & Bob's return Journey Bulbs
N0 transverse Doppler shift at Zero Momentum Rest Frame
All photons are emitted at c in all directions.
mass of bulb
$1/\sqrt{1-0^2/c^2} = 1.0$ kg
interval of clock
$1/\sqrt{1-0^2/c^2} = 1.0$ sec/day
photon wavelengths in all directions
$\lambda = 1.0$
Photon CBR Rest

Adam's Earth Rest Photon Bulb
Transverse photon Doppler shift $\lambda = 1.00000078$ m
Each photon is emitted at a different value for c +/- v relative to CBR.
$\lambda = 1.00125$, $\lambda = .99875$
mass of bulb
$1/\sqrt{1-v^2/c^2} = 1.00000078$ kg
interval of clock
$1/\sqrt{1-v^2/c^2} = 1.00000078$ sec
wavelength of photon
$\lambda = 1.00125$
$\lambda = .99875$
Earth Rest
375 km/s

Bob's outward & Chad's return Journey PhotonBulbs
Transverse photon Doppler shift $\lambda = 1.0000031$ m
Each photon is emitted at a different value for c +/- v relative to CBR
$\lambda = 1.0025$, $\lambda = .9975$
mass of bulb
$1/\sqrt{1-v^2/c^2} = 1.0000031$ kg
interval of clock
$1/\sqrt{1-v^2/c^2} = 1.0000031$ sec
wavelength of photon
$\lambda = 1.0025$
$\lambda = .9975$
750 km/s

Triplet Paradox Experimental Values

Clock interval of Zero Momentum Frame - $t' = T/\sqrt{1-0^2/c^2} = 1.0$
Lorentz mass of relative velocity ------- $m' = M/\sqrt{1-375^2/c^2} = 1.000000781$ kg
Clock interval of relative velocity --------- $t' = T/\sqrt{1-375^2/c^2} = 1.000000781$
Lorentz mass of fastest legs -------------- $m' = M/\sqrt{1-750^2/c^2} = 1.000003124$ kg
Lorentz mass of stationary clocks ---------- $m' = M/\sqrt{1-0^2/c^2} = 1.0$ kg
Clock interval of fastest legs -------------- $t' = T/\sqrt{1-750^2/c^2} = 1.000003124$
Clock interval of stationary "legs" ---------- $t' = T/\sqrt{1-0^2/c^2} = 1.0$
Average clock interval of to and fro legs ---------------------- $= 1.000001562$
Increased mass of Earth at 375 km/s -------------------------- $= 1.000000781$
Clock interval of Earth's 375 km/s Lorentz velocity frame -- $= 1.000000781$
Measured interval between Earth clock & Triplet's clock ---- $= 1.000000781$
Measured mass and time interval of Earth clock------- $m \& t' = 1.0$

Why Einstein Was an Ignorant Fool   *James Carter*

verse and Special Relativity's calculations must be based on the ZMF in order to yield correct values for experimental measurements of the time intervals in individual legs in Twin Paradox experiments.

Not even the most dedicated relativity enthusiast can deny that the location of the Living Universe's Zero Momentum Frame represents the preferred rest frame from which all accelerometer readings, Lorentz transformations, and photon Doppler shifts are ultimately calculated and measured. All photons move at exactly c within this frame and it is the only frame in which it is possible to make correct calculations for one-directional time dilations. It is also the only frame in which there are zero Doppler shifts in emitted and absorbed photons. While the very existence of the Doppler effect demands a common rest frame for photons, by their very nature, Doppler shifts can only be used to measure relative motion.

Some people who believe in intrinsic relative motion claim that only two-way relative motion time dilations can be measured. However, accurate one-way measurements of mass and time dilations are made in both the GPS clock rate calculations and the Pound-Rebka photon momentum and time measurements. Even true believers in aether frames will usually choose the ZMF as the location for their aether since it appears to be tied to the speed of light. They propose that all photons are measured to move at c relative to this zero velocity Lorentz aether.

Other aether people try to put their aether's rest frame at the same location as the $2.7°K$ Cosmic Blackbody Radiation's Lorentz frame. Special relativity does not attempt to locate positions of rest for its metaphysical assumption of a 4-dimensional spacetime continuum field. It is a postulate of the theory that the parameter of absolute space can't be measured. The physical assumption of the ZMF calculates and measures the position and magnitude of Earth's momentum and establishes values for its mass, space, time, and gravitational force and motion.

Photon Physics without Metaphysics

The corpuscular view of light cannot explain the celerity of light. This was Newton's theory of light, renewed by the photon of the relativistic theory. The celerity of the photon is regarded as a law of Nature. Relativists think it is perfectly futile to seek an explanation.

Jean de Climont

Photon Structure and Dynamics

Circlon Synchronicity is an experimental principle of the mass, space, time, and gravity of atoms and photons. It is based on the circlon shaped physical mass structures of electrons, protons and photons. Circlon synchronicity means that all electrons, protons, and photons in the Living Universe have identical parameters and are in perfect synchronicity with one another in time and space. All atoms have an individual relative momentum vector of $p = mv$ and all photons have an absolute momentum vector of $p = mc$ relative to the zero momentum photon rest frame.

Aether Theories

Einstein's postulate for the constant speed of light has led many theorists to believe that some form of universal medium such as an aether or field must be required for all photons to move at the same speed of light c within the same universal reference frame. This is the way sound waves move in air and water with their speed depending on the temperature and density of the medium.

Just because a photon has wave characteristics, does not mean it has to have a medium to travel through. The photon itself is its own moving mass medium. It travels through empty space at c and C on its own inertial momentum and angular momentum $p = mc$ & $I\omega = m\lambda C/2\pi$. Photons are made out of the same coil shaped mass structures as atoms.

Although Einstein claimed not to believe in aether as a carrier of photon waves, he invented a similar aether-like field substance called a spacetime continuum for his massless photons to travel through. Since then, Einstein's critics have invented many different unmeasured metaphysical types of aether and fields to transmit massless photon wave units through space in a process analogous to the way that sound waves are transmitted though air and solid matter. None of these theories is ever physically complete because there is never any experimental verification of the aether itself. We can measure many things about atoms and photons but no one has ever been able to measure anything about aether without claiming that all measurements are about aether.

Some dissident theorists propose separate aethers for photons and gravity, while others try to invent a single aether that does everything. The most commonly proposed aethers are either rigid or elastic solids. Others propose massless continuous fluids, while still others imagine their aether to be composed of countless, either stationary or rapidly moving unmeasured particles to which each theorist gives his own peculiar pet names. Some theorists even propose their aether to be composed of a stationary lattice of electron/positron dipoles that are packed into space, back to back, wall to wall, and treetop tall. Some claim them to have an overall universal density similar to a neutron star. Ordinary matter has the miraculous ability to pass through this dense material lattice without any resistance. Still others claim that these "positron/electron" dipoles are massless because they imagine that while electrons have mass, antimatter positrons must have "anti-mass" and when they combine together into a neutral dipole they are massless. One theorist even claims that the universe is filled with otherwise indefinable and undetectable "black boxes" that are filled with "fibers" that vibrate as photon waves pass through them. The one thing that all of these postulated aether, field and particle mediums have in common is that none has ever been detected by an experimental measurement.

Electric and magnetic fields have been measured, but always in connection with electrons and protons and never in connection with the empty void of space itself. Electric and magnetic fields are structural stationary photon-like extensions of electrons and protons and have nothing to do with aether.

Surprisingly, few of Einstein's critics have ever questioned his initial metaphysical assumption that the photon is a massless wave/particle duality that moves as a wave disturbance through the spacetime medium. The photon is considered by convention to be massless even though photons are measured to carry momentum, angular momentum, and energy as they travel through space. The only thing that we can ever measure about a massless photon is its momentum.

In contrast, if photons are allowed to have mass, they can travel and spin from one end of the universe to the other under the power of their own inertia. Photon dynamics can be understood from the simple results of experimental measurements and there is no need to invent elaborate metaphysical theories to understand how photons move through the void of empty space. They are just like rifle bullets that contain the separate energies in their moving (v) and spinning (V) mass. The kinetic energy of a rifle bullet is $e = mv^2/2 + mVr^2/2$.

Photon Doppler Shift Measurements

All photon measurements show that when atoms moving with different relative velocities emit photons, their motions have no effect on the velocity c that the photons travel through space. The atom's individual motion produces Doppler shifts in the emitted photon's momentum and wavelength but has no effect on its angular momentum or its constant velocity c that all photons move through empty momentum space.

Since all photons are Doppler shifted by the absolute motion of the atoms that emit or reflect them and Doppler shifted again by the absolute motion of the device that measures them, there is no experimental way to determine the separate absolute motions of either source or observer. However, the sum of these Doppler shifts can always be used to measure the precise relative velocity between them. Also, even though the photon's relative velocity between source and observer will almost always be less than or greater than c, the average two way velocity between them will always be measured at exactly c.

Each photon measurement contains a pair of unknown Doppler shifts that can only be resolved as a single relative motion. While one way measurements of photon velocity v are always Doppler shifted at c+/-v, two way measurements of photon velocity cancel all Doppler effects and always show v at exactly $c = (c+v) + (c-v)$. One way measurements of photons show their relative motion with matter and two-way measurements show their absolute motion with other photons.

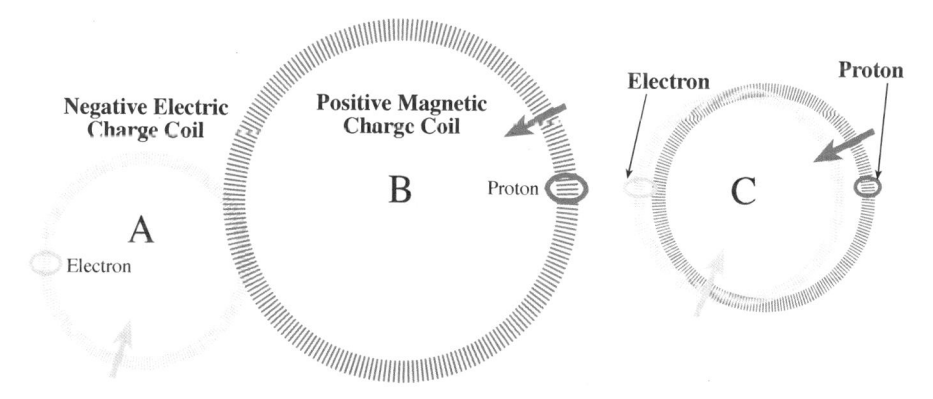

Charge Coil Dynamics and Photon Emission

What theorists call electric and magnetic fields are actually structural parts of the electron and proton that extend outward from their edges and are not "fields" existing as separate entities within space. Electric and magnetic fields are large etherial particles extending out into space from each electron and proton respectively.

When the expanding negative electric charge coil of an electron (A) comes in contact with a proton's positive magnetic charge coil (B), they intertwine with one another and pull the two particles together (C).

As these two charge coils pull together (C), they align and adjust to form a stationary photon that reaches synchronicity when they become the same size and occupy the same space while spinning in opposite directions (D). The stationary photon's secondary coils spin in opposite directions while maintaining their centers on a single two-dimensional plane and the 11.7 times smaller primary coils spin in opposite directions with their centers on an infinite number of different two-dimensional planes.

D

| Secondary Coil Spin @ C | | Primary Coil Spin @ C |

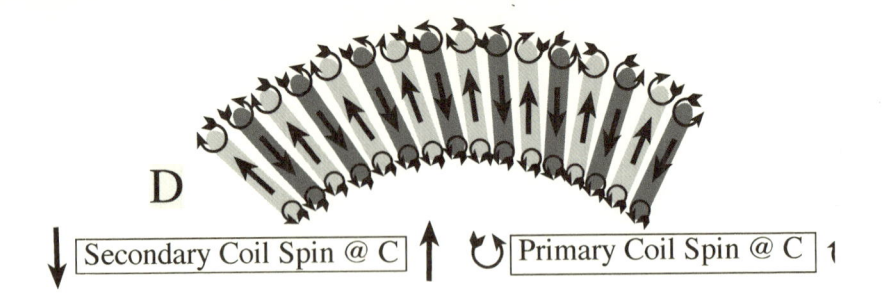

These two equal and opposite charge coils combine to become a stationary photon (E). It is the stationary photon link at the Bohr radius that holds the proton (F) and electron (G) together within the hydrogen atom.

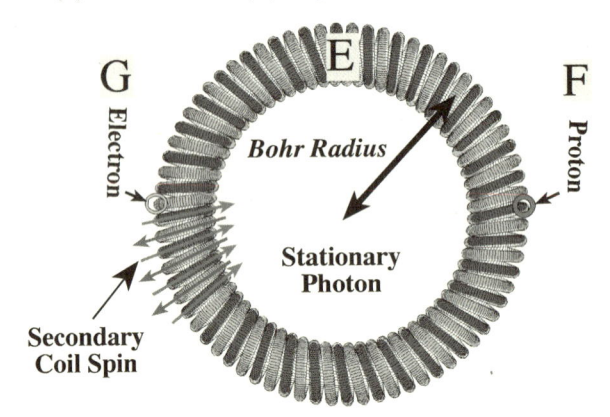

G — Electron **E** **F** — Proton

Bohr Radius

Stationary Photon

Secondary Coil Spin

When this alignment process reaches synchronicity, the negative electric coil structure of the electron and the positive magnetic coil structure of the proton combine to form a stationary electric/magnetic photon. This photon has a mass = 2, a wavelength = 1/2 and an angular momentum of $h/\pi = m\lambda c/\pi$. It has an energy of $e = mc^2 = 2$, Half of this is the rotational kinetic energy $= mC^2/2 = 1$ from the primary coils spinning at C on all planes and the other half is from the rotational kinetic energy $= mC^2/2 = 1$ of the secondary coils spinning in opposite directions at the speed at C with their centers on a single plane. The tertiary coils move in one dimension, the secondary coils move in two dimensions and the primary coils move in n dimensions.

The circumference of the Bohr link stationary photon is $2\pi a_o$ and when it splits into a pair photon they each have a wavelength of $4\pi a_o/\alpha$. This is $2/\alpha$ times longer than the Bohr link's circumference.

When a stationary photon bifurcates, the oppositely spinning magnetic and electric secondary coils break in two and separate their two spins at C into two opposite directions at c. As the two new photons unravel from the Bohr link's secondary coils, they stretch out to twice the secondary coils original length. The emitted photons are the stretched out lengths of oppositely spinning primary coils.

Why Einstein Was an Ignorant Fool *James Carter*

Internal Coil Structure of the Hydrogen Atom

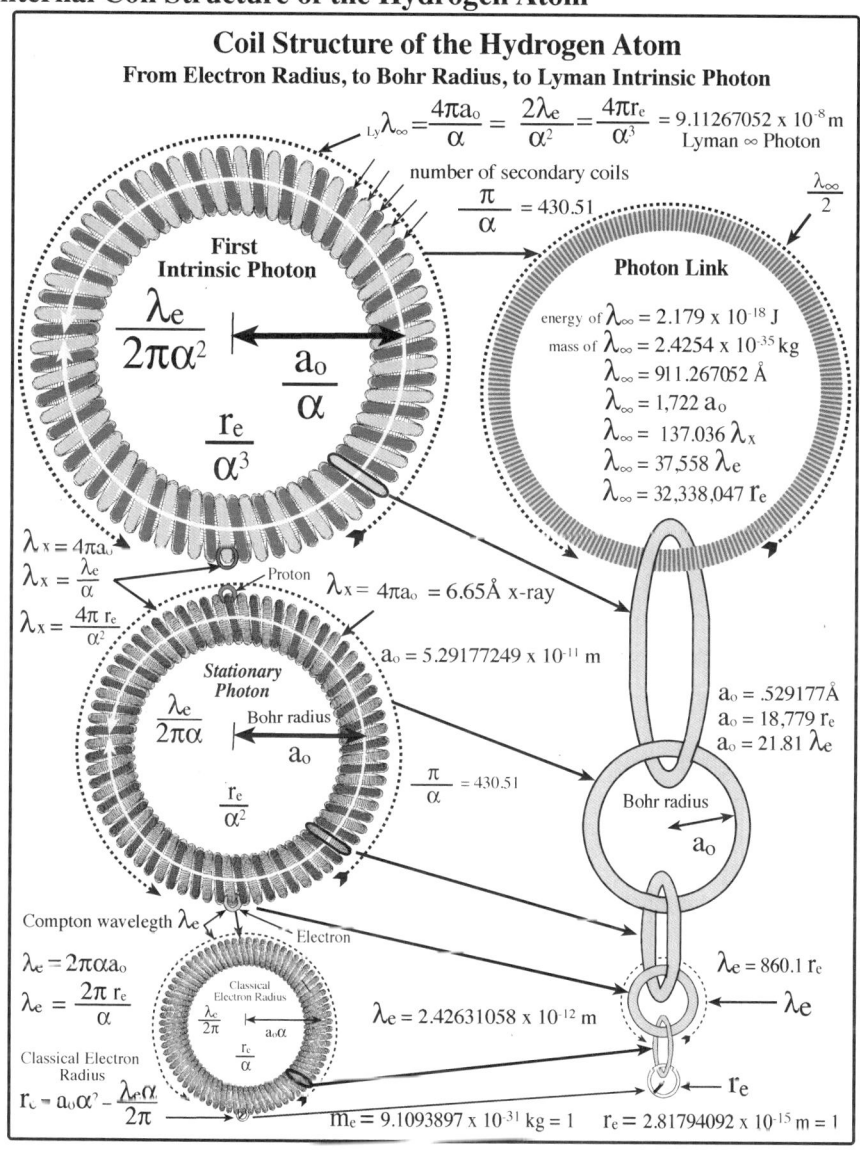

Coil Structure of the Hydrogen Atom
From Electron Radius, to Bohr Radius, to Lyman Intrinsic Photon

$$_{Ly}\lambda_\infty = \frac{4\pi a_o}{\alpha} = \frac{2\lambda_e}{\alpha^2} = \frac{4\pi r_e}{\alpha^3} = 9.11267052 \times 10^{-8} \text{ m}$$
Lyman ∞ Photon

number of secondary coils
$$\frac{\pi}{\alpha} = 430.51$$

$$\frac{\lambda_\infty}{2}$$

First Intrinsic Photon

$$\frac{\lambda_e}{2\pi\alpha^2} \quad \frac{a_o}{\alpha}$$

$$\frac{r_e}{\alpha^3}$$

Photon Link

energy of $\lambda_\infty = 2.179 \times 10^{-18}$ J
mass of $\lambda_\infty = 2.4254 \times 10^{-35}$ kg
$\lambda_\infty = 911.267052$ Å
$\lambda_\infty = 1,722\ a_o$
$\lambda_\infty = 137.036\ \lambda_x$
$\lambda_\infty = 37,558\ \lambda_e$
$\lambda_\infty = 32,338,047\ r_e$

$\lambda_x = 4\pi a_o$
$\lambda_x = \frac{\lambda_e}{\alpha}$
$\lambda_x = \frac{4\pi r_e}{\alpha^2}$

Proton $\lambda_x = 4\pi a_o = 6.65$ Å x-ray

$a_o = 5.29177249 \times 10^{-11}$ m

Stationary Photon

$$\frac{\lambda_e}{2\pi\alpha} \quad \text{Bohr radius} \quad a_o$$

$$\frac{r_e}{\alpha^2}$$

$$\frac{\pi}{\alpha} = 430.51$$

$a_o = .529177$ Å
$a_o = 18,779\ r_e$
$a_o = 21.81\ \lambda_e$

Bohr radius a_o

Compton wavelength λ_e — Electron

$\lambda_e = 2\pi\alpha a_o$
$\lambda_e = \frac{2\pi r_e}{\alpha}$

Classical Electron Radius
$\frac{\lambda_e}{2\pi} \quad a_o\alpha$
$\frac{r_e}{\alpha}$

$\lambda_e = 2.42631058 \times 10^{-12}$ m

$\lambda_e = 860.1\ r_e$ ← λ_e

Classical Electron Radius
$r_e = a_o\alpha^2 = \frac{\lambda_e\alpha}{2\pi}$

← r_e

$m_e = 9.1093897 \times 10^{-31}$ kg = 1 $r_e = 2.81794092 \times 10^{-15}$ m = 1

It is quite difficult make realistic drawings depicting the mechanics of the hydrogen atom because of the vast size differences between the different links in its radiation chain. Of the 7 links shown here, the photon link is the largest link in the chain and is over $\sqrt{\alpha^7} = 32,000,000$ times larger than the electron's smallest classical electron radius link. Each consecutive link is $\sqrt{\alpha} = 11.7$ times larger than the previous link.

The above drawings show two different ways of depicting the Hydrogen atom's circlon shaped structural links. From the outside, the whole chain appears as just its largest link because the progressively smaller links are hidden inside of the atom's seven coils.

Why Einstein Was an Ignorant Fool — *James Carter*

Photon Energy & Momentum

$E = mc^2/2 + mC^2/2 = mcC$ & $p = mc$

PHOTON

- Positive Magnetic Matter from Proton
- Negative Electric Matter from Electron

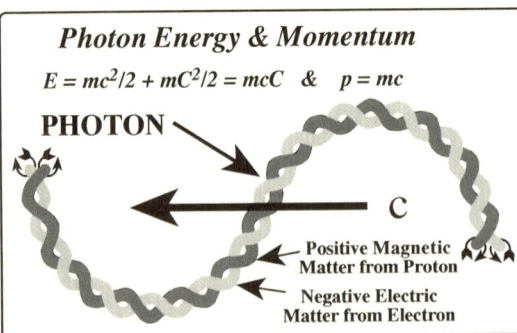

Photon Angular Momentum = $M\lambda C/2\pi$
Photon Momentum = Mc

The photon's angular momentum is contained in the wave-like motion of the two oppositely spinning bodies of positive magnetic matter and negative electric matter. The opposite spins at C of this positive and negative hollow string contain the photon's absolute rotational energy and the kinetic energy contained in the photon's linear momentum is relative to an observer's absolute motion.

Photon Dynamics

When synchronicity is reached within the stationary photon structure (E), it splits into two photons that have equal wavelengths, equal and opposite momenta and equal and opposite angular momenta on opposing planes. Each photon has $m = 1$, $\lambda = 1$, $I\omega = h/2\pi = m\lambda c/2\pi$ and $e = 1$. This energy is one half the kinetic energy $e = mc^2/2$ of the photon's vector at c and the other half is the rotational energy of its primary coils spinning at C on all planes and in opposite directions. $e = mc^2/2 + mC^2/2 = mcC$. When an atom emits a photon, there is no transformation between mass and energy. What happens is the transformation of two opposite rotational motions of mass at C into two opposite linear motions of mass at c. A quantity of angular momentum becomes two equal quantities of linear momentum.

The random relative motions (heat) of the electron and proton prior to the coupling of their charge coils produce the atom's initial quantity of angular momentum $I\omega = mvr$ (A,B and C arrows). The number of $I\omega = h/2\pi$ units determines the number of photons that the atom can emit until its angular momentum becomes less than $2h/2\pi = h/\pi$ and it does not have enough $I\omega$ to emit a pair of photons. The atom reaches its ground state when it can no longer emit any photons until it receives one or more units of angular momentum from another photon or from random mechanical motion (heat).

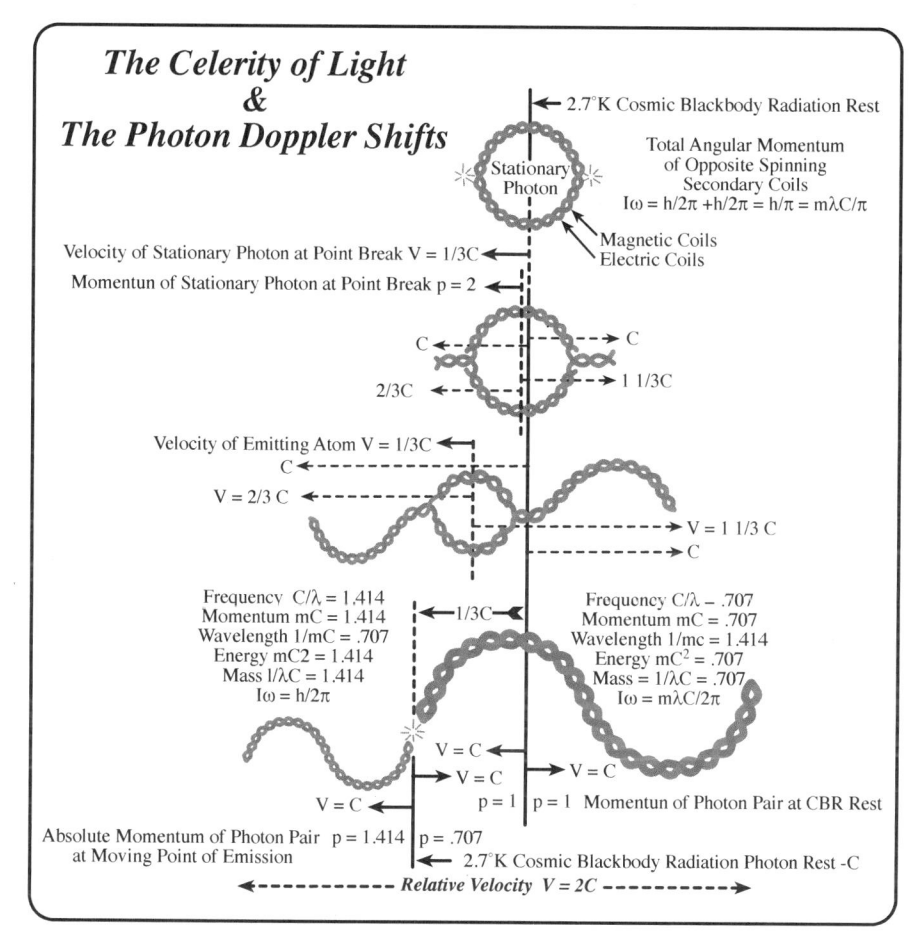

The Celerity of Light & The Photon Doppler Shifts

← 2.7°K Cosmic Blackbody Radiation Rest

Stationary Photon

Total Angular Momentum of Opposite Spinning Secondary Coils
$I\omega = h/2\pi + h/2\pi = h/\pi = m\lambda C/\pi$

Magnetic Coils
Electric Coils

Velocity of Stationary Photon at Point Break V = 1/3C
Momentun of Stationary Photon at Point Break p = 2

C ← → C
2/3C ← → 1 1/3C

Velocity of Emitting Atom V = 1/3C
C
V = 2/3 C
V = 1 1/3 C
C

Frequency $C/\lambda = 1.414$
Momentum $mC = 1.414$
Wavelength $1/mC = .707$
Energy $mC2 = 1.414$
Mass $1/\lambda C = 1.414$
$I\omega = h/2\pi$

←1/3C←

Frequency $C/\lambda = .707$
Momentum $mC = .707$
Wavelength $1/mc = 1.414$
Energy $mC^2 = .707$
Mass $= 1/\lambda C = .707$
$I\omega = m\lambda C/2\pi$

V = C
V = C → V = C
V = C   p = 1 | p = 1 Momentun of Photon Pair at CBR Rest

Absolute Momentum of Photon Pair p = 1.414 | p = .707
at Moving Point of Emission

← 2.7°K Cosmic Blackbody Radiation Photon Rest -C

←--------- Relative Velocity V = 2C ---------→

The Mechanics of Photon Motion C+ & C-

All photons are emitted as identical pairs from the common position of photon zero momentum rest. The stationary photon has zero net energy because the opposite angular momenta of its coils cancel and there is no relative energy between them. Individually, the coils have energy but together their energies cancel.

Within the stationary photon, the magnetic coils of the proton and the electric coils of the electron are spinning at C and in opposite directions while adjusting and aligning with one another. When they reach reverse synchronicity, both coils bifurcate and combine into two identical photons that move apart on a single vector at 2 c.

When the coils divide, the secondary magnetic coils from the proton combine with the opposite secondary electric coils from the electron. They stretch out into the electric/magnetic primary coil wavelength. Photons are $2/\alpha = 274$ times larger than the stationary photon that emitted them. This is because when the secondary coils of a circlon shape are stretched out into a primary coil wavelength λ, they are $2/\alpha$ times longer than the circumference of the stationary photon's tertiary coil. Opposite halves of the electric/magnetic coils combine to form a pair of identical photons moving in opposite directions.

Why Einstein Was an Ignorant Fool   *James Carter*

When photons are emitted from a moving atom they share momentum with it but not velocity. During the time f the stationary photon is being transformed into a pair of photons, their wavelengths and momenta are being Doppler shifted by the atom's motion.

Even though these photons are identical at the point of photon break, they become red and blue shifted during the time of emission by sharing momentum with the atom. In the time λ/c between photon break and emission, momentum is added to one photon and removed from the other. Photons share momentum with the moving atom but not its velocity relative to zero momentum rest. Photons get all of their velocity c from the opposite spins of the electric/magnetic secondary coils of the stationary photon. Photons get no velocity from the emitting atom's linear motion. All of a photon's velocity comes from the secondary coil spin velocities of the stationary photon C. The velocity c of photons is always constant because they get all of their velocity from the coils of matter spinning at C and none from the inertial motions v of matter. The absorption of a photon by an atom is virtually the same process in reverse.

The 2.7° CBR Dipole Anisotropy

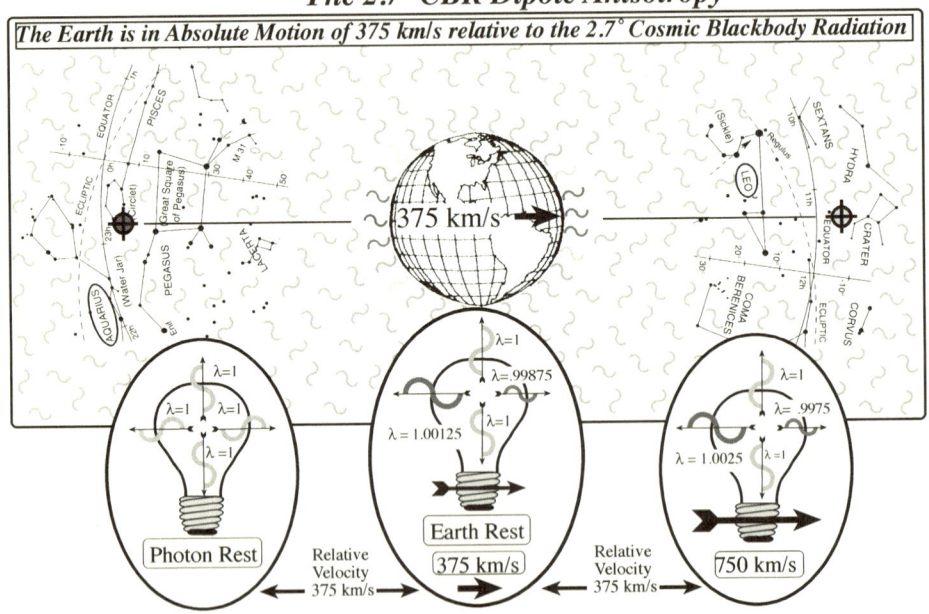

The Earth is in Absolute Motion of 375 km/s relative to the 2.7° Cosmic Blackbody Radiation

This drawing shows the unseen Doppler shifts caused by Earth's absolute motion toward Leo relative to the Zero Momentum Rest Frame of the Cosmic Blackbody Radiation as measured by the CBR dipole anisotropy. A moving light bulb emits red and blue Doppler shifted photons on opposite sides of the bulb along the vector of motion.

Photon Flywheel

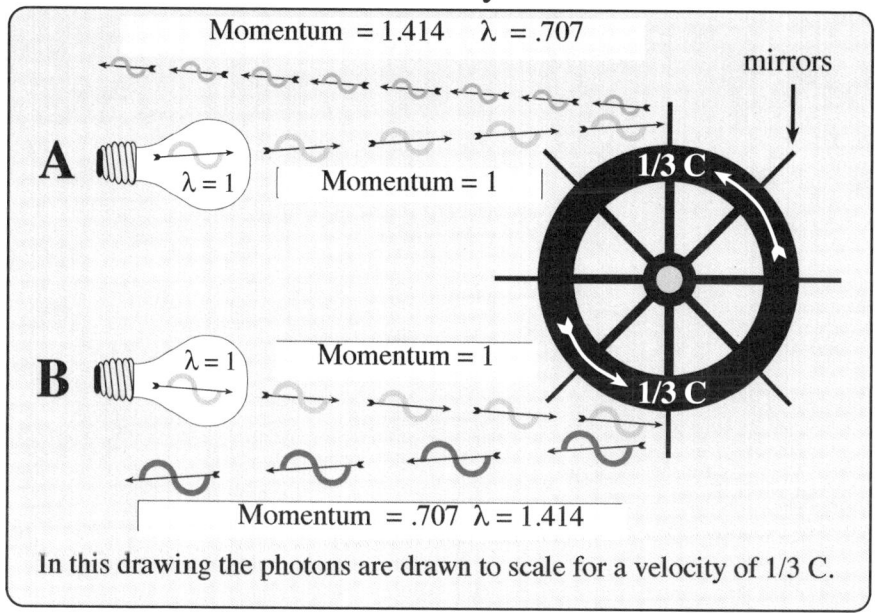

In this drawing the photons are drawn to scale for a velocity of 1/3 C.

Top beam: Momentum = 1.414, λ = .707
Laser A: λ = 1, Momentum = 1
Wheel mirrors at 1/3 C
Laser B: λ = 1, Momentum = 1
Bottom beam: Momentum = .707, λ = 1.414

    For an example of how the change in a body's kinetic energy must also change its mass, consider a thought experiment in which a flywheel has evenly spaced mirrors attached to its outer surface like the fins of a paddle-wheel. The wheel is made of an exceedingly strong imaginary material and is spun so fast that the mirrors are moving at a velocity of 1/3 c.

    Two lasers, A and B, shoot photons at the mirrors on opposite sides of the wheel so that the mirrors are moving at 1/3 c toward the photons from laser A and at 1/3 c away from laser B photons. These photons are all emitted from the lasers with a wavelength, and momentum of exactly one, and all move at exactly c relative the same inertial CBR rest frame common to all photons. These photons reflect from the mirrors at the same velocity c that they had before striking the mirror. The velocity of the mirrors has no effect on the photons' velocity but does change their momentum and wavelength.

    The λ = 1 photons from laser A are blue-shifted to a wavelength of λ = .707 as they reflect from the approaching mirror, and their energy and mass are increased to 1.414. In this process, the velocity of the spinning wheel is slowed as mass and energy are transferred to the reflecting photons.

    The photons from laser B are red-shifted as they reflect from the receding mirror to a wavelength of λ = 1.414 and a momentum of .707. In this case, the velocity of the wheel increases as energy/mass is transferred from the photons to the wheel. In both of these examples, momentum is conserved and both mass and energy remain separate and constant. Mass and energy are two sides of the same coin and always remain constant, conserved and equal.

    If we attempt to explain this experiment in terms of massless photons then the conservation of mass and energy is lost. The photons from laser A take energy away from the wheel and decrease its mass. Laser B photons transfer energy to the wheel and increase its mass. In both cases, energy remains constant but mass either vanishes into or appears from nowhere. How can mass and energy be equivalent if energy remains constant but mass does not? If the energy of moving mass produces extra mass how can the energy of moving photons not have mass?

Why Einstein Was an Ignorant Fool   *James Carter*

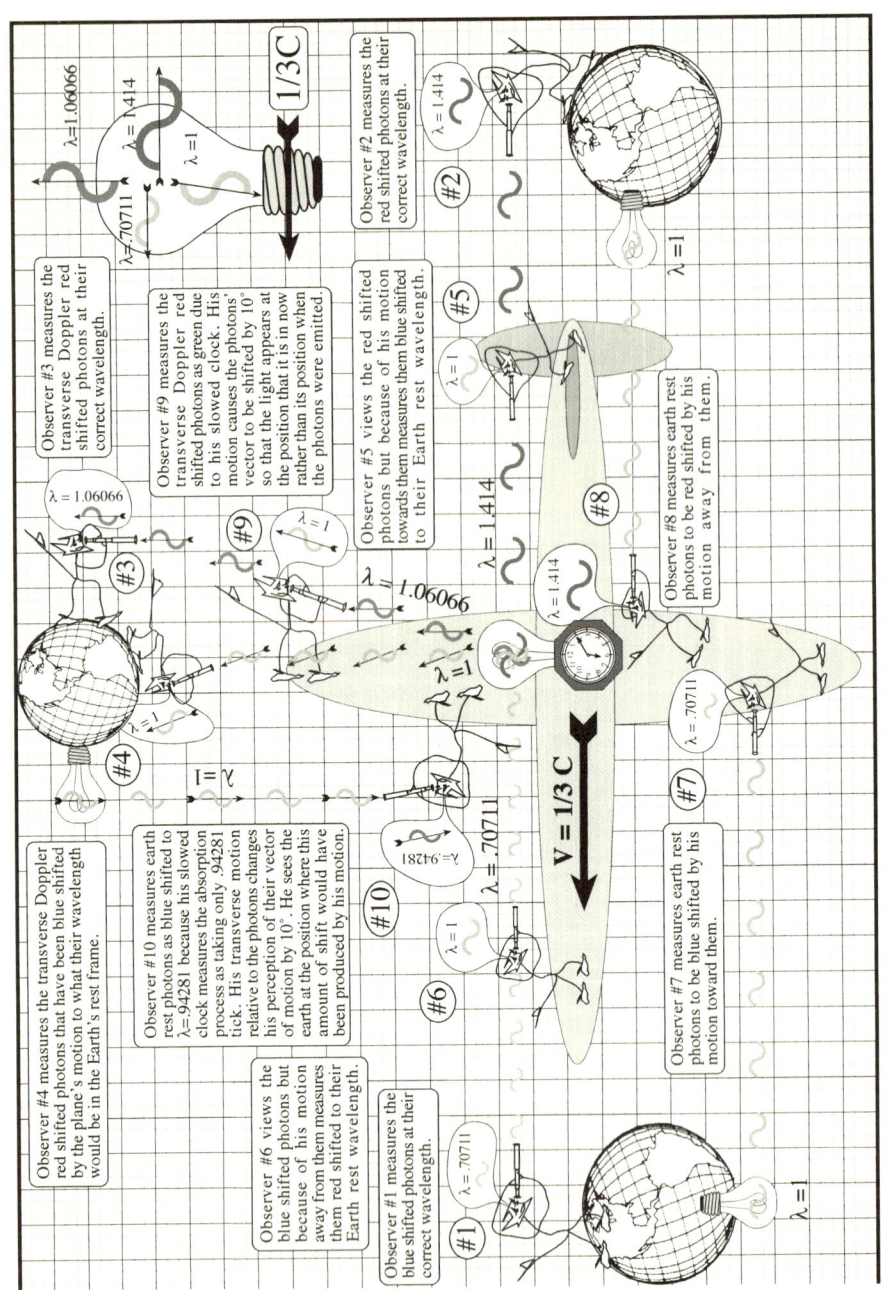

This is a scale model of photons being Doppler shifted in ten different situations by one third of the speed of light. Photons emitted with the motion of the plane have half the wavelengths and twice the energy of the photons emitted against its motion.

The above drawings all reproduce experimental measurements of photon force and motion and do not involve any "theories" of photons proposing aethers, fields, dimensions and any other unmeasured parameters or metaphysical media.

Binary Pulsar Observations

A B C D A B C D A B C D A B C D A B C D A B C D A B

The photons from this pulsar have traveled for two hundred thousand years before being observed and measured on earth, yet to the limit of measurement they remain in perfect order of the time they were emitted. They form a line of photons with wavelengths that smoothly change back and forth from between ($\lambda = 1.001$) and ($\lambda = .999$) as the pulsar revolves around its companion.

1,001.1 photons/sec
999.1 Photons/sec
1,000.1 Photons/sec

30 km/sec = .0001C
Winter
Relative velocity between photon and observer = 1.0001C

$\lambda = .999$
300 km/sec = .001C
$\lambda = 1$
1,000 Photons/sec
$\lambda = 1.001$
300 km/sec
$\lambda = 1.001$

Sun

Spring ↓
Relative velocity between photon and observer = 1.0000C

Relative velocity between photon and observer = .9999C
Summer
30 km/sec = .0001C

999.9 Photons/sec
1000.9 Photons/sec
998.9 Photons/sec

When the observer uses the 1000 pulses per second of the pulsar as a measure of time to measure the velocity of the photons relative to his moving observation point on Earth, he finds that the measured velocity of light is not always c in his rest frame. In spring, when the earth's orbital motion is transverse to the pulsar's position, the green photons from pulsar positions (A) and (B) are measured at 1000 pulses/sec. In summer, when he is traveling at .0001c away from the pulsar, he receives the green photons at 999.9 pulses/sec and thus measures the velocity of the photons to be .9999c in his rest frame. In winter, when he is traveling at .0001c toward the pulsar he receives 1.001 pulses/sec and thus measures the velocity of the photons to be 1.001c.

◄──────── 200,000 light years ────────►

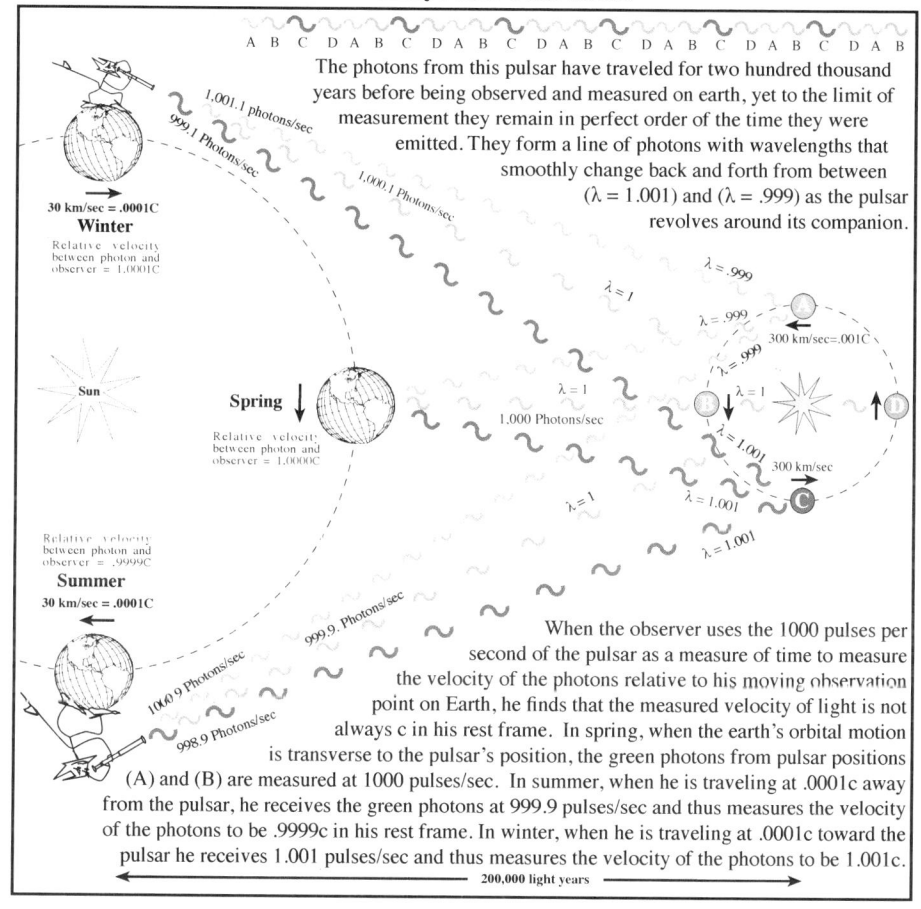

Binary Pulsars

The observation of binary pulsars offers very convincing experimental evidence that all photons move at exactly c within the common reference frame of Zero Momentum photon rest. A binary pulsar emits rapid bursts of X-ray photons at very regular intervals as it revolves around a companion star. When photons from a pulsar are carefully measured, it is found that they are blue shifted when the revolving pulsar is moving toward the earth and red shifted when the pulsar is moving away. Even though the pulsar may be two hundred thousand light years from earth, the photons remain perfectly lined up in their order of emission. They are observed as repeating sequences of first red shifted photons and then blue shifted photons. If the changing motion of the revolving pulsar had any effect on the photons' velocity of c, then the photons could never have remained in their sequence of emission for two hundred thousand years. If any of these photons moved even slightly faster or slower than c, they would be observed as a jumbled up mixture of red and blue shifted photons.

Why Einstein Was an Ignorant Fool   *James Carter*

Circlon Coil Spin of Electrons & Protons

4 Photon Energies
Rotational Kinetic Energy
$e = mC^2/4 + mC^2/4 = mCC/2$
$+$
Linear Kinetic Energy
$e = mc^2/4 + mc^2/4 = mcc/2$
$e = mcC$

Linear Motion @ c
Opposite Photon Coil Spin @ C
Positive Magnetic Coils
Negative Electric Coils
Tertiary Coil

Secondary Coil Spin   Primary Coil Spin

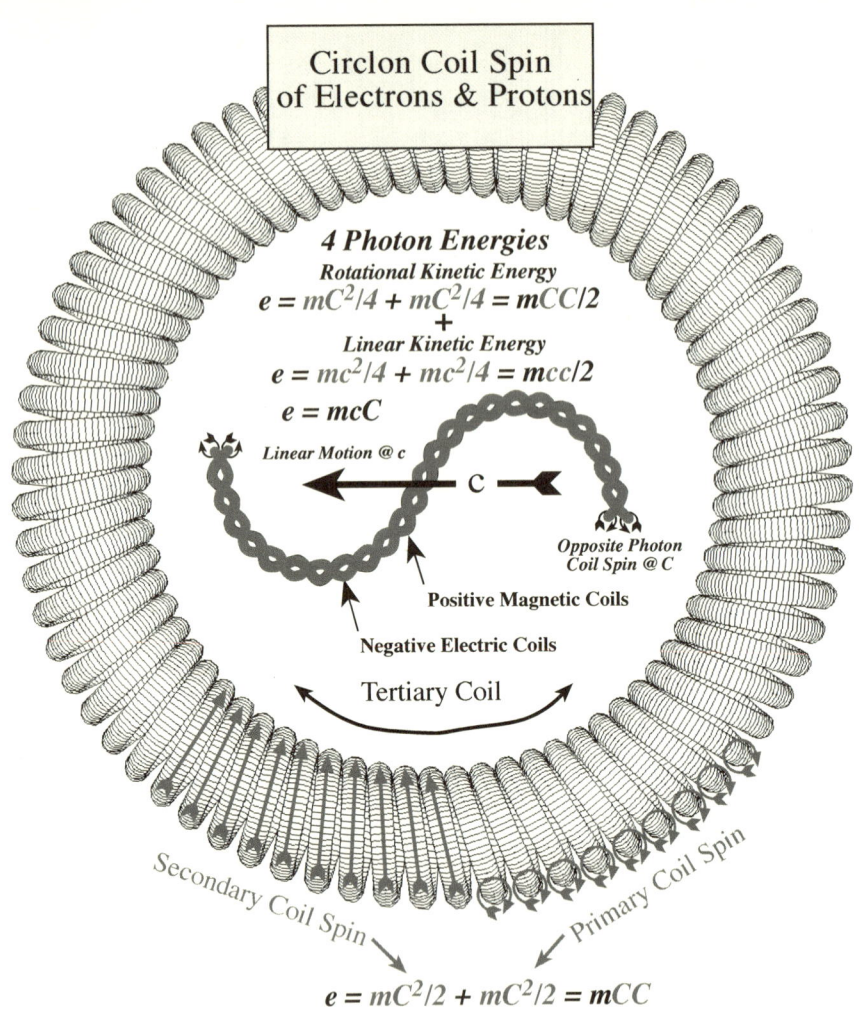

$e = mC^2/2 + mC^2/2 = mCC$
Proton & Electron Rest Mass/Energy

The True Meaning of E = MC²

The energy of the photon is divided into four separate but equal kinetic energies. The linear energy of its magnetic coil mass, the linear energy of the electric coil mass, the rotational energy of the magnetic coil mass, and the rotational energy of the electric coil mass. The two linear coil energies are Doppler shifted by the relative motion of the observer but the two opposite rotational energies of the electric/magnetic coils are absolute and are measured to be the same in all moving frames.

The two separate but equal rest mass rotational energies of the primary and secondary coils of the electron and proton are absolute and measured to be the same by all moving observers. When electrons and protons are accelerated, their momentum and energy is measured as an increase in mass but this "relativistic mass and energy" is separate from the particles' rest mass/energy.

Why Einstein Was an Ignorant Fool   *James Carter*

Experimental Measurement of Photon Mass

Doppler shifts in photon momentum and energy could be detected with lasers aligned to the 375 km/s velocity vector of the CBR. Such an experiment could easily be performed in a small underground laboratory. Four lasers would be aligned with, against, and at right angles to the CBR dipole anisotropy. Sensitive detectors would measure changes in the photon's momentum and energy as Earth's rotation moved the lasers into and out of alignment with the vector of CBR motion.

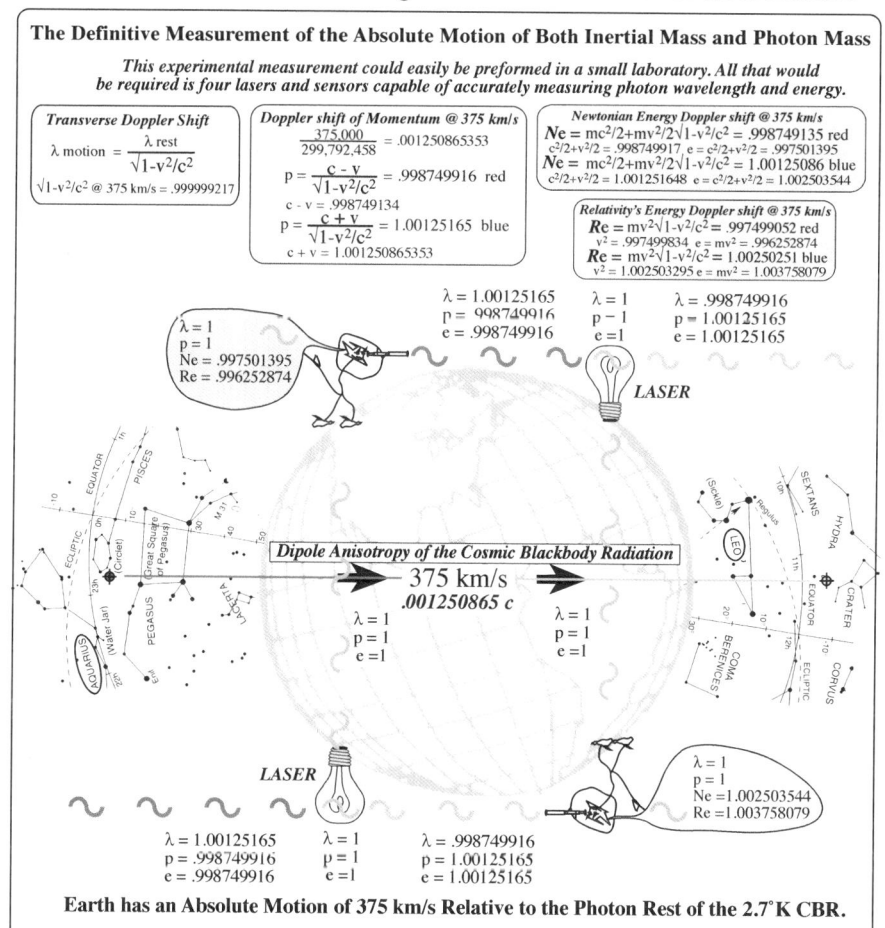

The Definitive Measurement of the Absolute Motion of Both Inertial Mass and Photon Mass

This experimental measurement could easily be preformed in a small laboratory. All that would be required is four lasers and sensors capable of accurately measuring photon wavelength and energy.

Transverse Doppler Shift

$$\lambda_{motion} = \frac{\lambda_{rest}}{\sqrt{1-v^2/c^2}}$$

$\sqrt{1-v^2/c^2}$ @ 375 km/s = .999999217

Doppler shift of Momentum @ 375 km/s

$$\frac{375,000}{299,792,458} = .001250865353$$

$$p = \frac{c-v}{\sqrt{1-v^2/c^2}} = .998749916 \text{ red}$$

$c - v = .998749134$

$$p = \frac{c+v}{\sqrt{1-v^2/c^2}} = 1.00125165 \text{ blue}$$

$c + v = 1.001250865353$

Newtonian Energy Doppler shift @ 375 km/s

$Ne = mc^2/2 + mv^2/2\sqrt{1-v^2/c^2} = .998749135$ red
$c^2/2 + v^2/2 = .998749917$, $e = c^2/2 + v^2/2 = .997501395$
$Ne = mc^2/2 + mv^2/2\sqrt{1-v^2/c^2} = 1.00125086$ blue
$c^2/2 + v^2/2 = 1.001251648$ $e = c^2/2 + v^2/2 = 1.002503544$

Relativity's Energy Doppler shift @ 375 km/s

$Re = mv^2\sqrt{1-v^2/c^2} = .997499052$ red
$v^2 = .997499834$ $e = mv^2 = .996252874$
$Re = mv^2\sqrt{1-v^2/c^2} = 1.00250251$ blue
$v^2 = 1.002503295$ $e = mv^2 = 1.003758079$

$\lambda = 1.00125165$ $\lambda = 1$ $\lambda = .998749916$
$p = .998749916$ $p = 1$ $p = 1.00125165$
$e = .998749916$ $e = 1$ $e = 1.00125165$

$\lambda = 1$
$p = 1$
$Ne = .997501395$
$Re = .996252874$

LASER

Dipole Anisotropy of the Cosmic Blackbody Radiation
375 km/s
.001250865 c

$\lambda = 1$ $\lambda = 1$
$p = 1$ $p = 1$
$e = 1$ $e = 1$

$\lambda = 1$
$p = 1$
$Ne = 1.002503544$
$Re = 1.003758079$

LASER

$\lambda = 1.00125165$ $\lambda = 1$ $\lambda = .998749916$
$p = .998749916$ $p = 1$ $p = 1.00125165$
$e = .998749916$ $e = 1$ $e = 1.00125165$

Earth has an Absolute Motion of 375 km/s Relative to the Photon Rest of the 2.7° K CBR.

This motion can be detected independently of the CBR by measureing the difference in Doppler shifts between a photon's energy and its wavelength and momentum. This measurement can also be used to show the photon has a mass structure and that it is not just "pure energy".

This experiment measures a photon's Doppler shifted momentum, energy and wavelength. It will show a different value between Einstein's idea of the "pure energy" massless photon and the physical parameters of mass, space, and time in the measurements of photon momentum, wavelength, and energy. These experiments identify photon mass but are unable to separate its value from the unknown Doppler shift in a photon's absolute momentum. Einstein's massless photon would have the same proportional photon Doppler shifts for momentum and energy. A photon with mass would have no Doppler shifts in momentum or wavelength a 50% less Doppler shift in its energy..

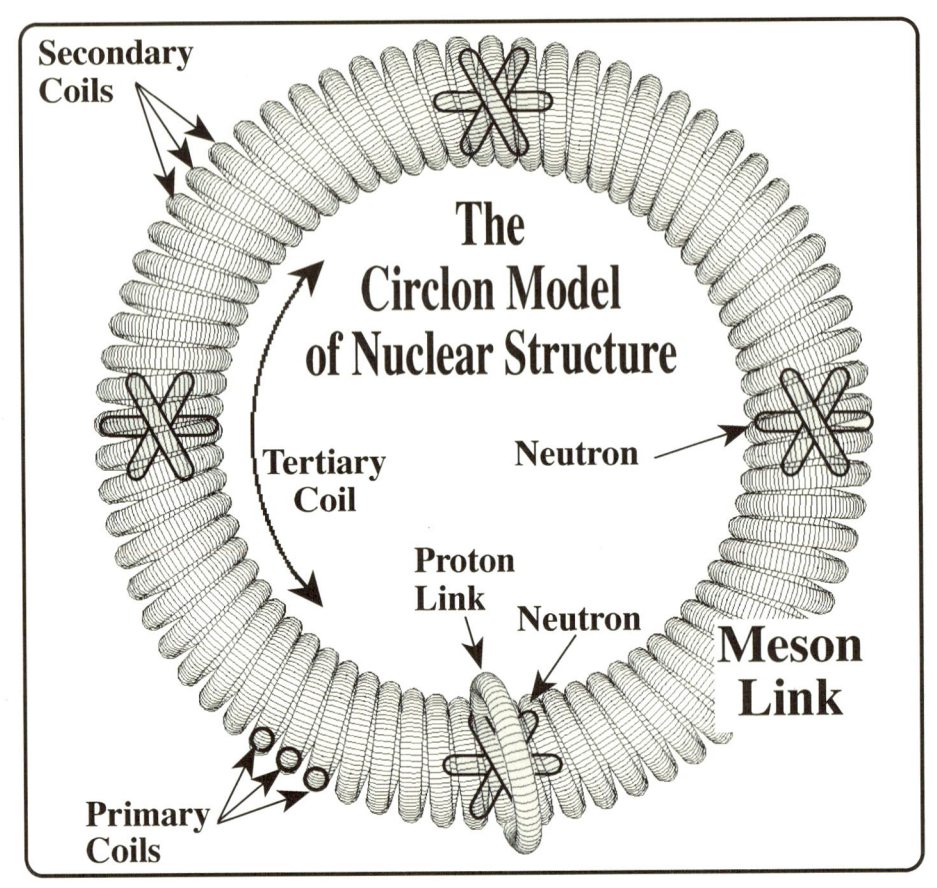

The Circlon Model of Nuclear Structure

(Diagram labels: Secondary Coils; Tertiary Coil; Neutron; Proton Link; Neutron; Meson Link; Primary Coils; The Circlon Model of Nuclear Structure)

The Circlon Model of Nuclear Structure

The Circlon Shape
All particles of matter, are combinations or configurations of four basic stable particles. These are protons, electrons, photons, and neutrinos. All of these particles can be created when other particles of matter collide with one another at velocities near the speed of light. Each new particle is always created with an exact opposite antiparticle. These are the antiproton, positron, photon, and antineutrino. All eight of these particles are constructions of two basic kinds of hollow string that have mass, dimension, and shape. Particles with positive charge like protons and positrons are composed of positive magnetic string and particles with a negative charge like electrons and antiprotons are composed of negative electric string. Each photon is composed of an equal piece of both electric and magnetic string. Neutrinos are composed of a piece of magnetic string and antineutrinos are made of electric string.

To form the physical structure of each particle, this cosmic string is wound into several series of different sized coils that form a structure called a circlon shape. The circlon has the basic shape of a torus that is composed of several series of smaller structures with the circlon shape. The circlon has a *tertiary coil* that is composed of smaller circular-shaped *secondary coils* that are composed of smaller circular shaped *primary coils*. At this point we may assume that these coils are composed of a hollow string, which is the fundamental component of reality.

The circlon shape is fundamental to particles of matter in the universe. It exists in essentially two varieties; the proton (positive string) and the electron (negative string). These are identical except for oppositeness in internal spins (charge) and a difference in scale. The electron is 1,836 times larger than the proton and the proton is 1,836 times more massive than the electron.

The Electromagnetic Charge Chain

Protons and electrons are totally mechanical particles of matter that must touch to interact. It was long considered that each particle was attached to its own electromagnetic field that could be extended far out into the space around them. Instead of fields, the circlon shaped protons and electrons have potential sizes that are unlimited in scope. Once formed, protons and electrons immediately extend their size by creating a single chain of progressively larger links with the proton or electron at one end and the direction of infinity at the other. These chains never get close to infinite length because they soon interact with the single chain of another proton or electron. As soon as a proton and electron connect, their chains switch from expansion to contraction and they are pulled together. As the links in the chain get progressively smaller, the last links at the end of each chain combine to emit a pair of photons as the next smaller links in the chain connect. As the atom gets smaller and smaller it emits photons of shorter and shorter wavelengths. Each time an atom emits a photon it gives up a unit of angular momentum ($I\omega = h/2\pi = m\lambda c/2\pi$). An atom stops emitting photons when the angular momentum between the proton and electron is less than the one unit needed to produce a photon. At this ground state, the charge chains are at equilibrium and the proton and electron maintain a constant distance. If angular momentum is added to a ground state atom the largest links in its chain will combine to produce photons. The charge chain of the proton has been called the "magnetic field" and the electron's charge chain is referred to as the "electric field".

The proton link is the first link in the proton's chain and contains most of its mass and maintains a position at the far edge of the circlon shaped particle. The meson link is the second link of the proton's chain and within an atom, its size and mass are ($1/\sqrt{\alpha}$) 11.7 times larger and 11.7 times less massive. The mass of a proton is 938 meV and the mass of a free pi-meson is 139.57 meV (938/139.57= 6.721).

Why Einstein Was an Ignorant Fool   *James Carter*

A meson link forms spontaneously from within the circlon structure of a bare proton. As soon as the meson link is formed, a third muon link is formed from within the structure of the meson link. An unlimited number of ever larger links will continue to form until the charge chain comes in contact with the charge chain of another particle.

The Meson Link

When protons are bombarded with high energy particles, the meson link can be broken free to become a pi-meson. The pi-meson is more massive than a meson link because it acquired mass from the large amount of energy that it took to break it loose from the proton link. The pi-meson is a well known particle with a relativity long lifetime. It decays when this bare meson link spontaneously forms a second link that transforms it into a muon. A meson is simply a physical part of the proton. Its structure grew from within one of the bare proton's secondary coils. It grows into the next fractal layer of the positive mass string that makes up the circlon shaped mass structure of a proton.

The Neutron

A neutron is basically a Hydrogen atom that is turned inside out. When the bond between a proton and electron are subjected to enough energy, their mutual charge chains collapse to the point where the bare electron becomes trapped inside of the bare proton link. Where the circlon shape of the proton was spinning like a wheel the neutron is now spinning end over end more like a ball. In this condition, neither particle can extend a charge chain out to other particles. The neutron's extra mass comes from the energy that it took to force the electron inside of the proton's structure. This extra mass makes the neutron smaller than a proton. Without a meson link, the neutron is just the right size to fit inside of the secondary coils of another proton's meson link. This forms a deuterium nucleus. A free neutron will decay within about nineteen minutes. When this happens, a proton and electron are not "created". They are simply released from the bond that they shared within the neutron. The extra mass of the neutron is released as the decay energy of the particles.

The meson link is a hollow torus shaped physical structure. It holds the neutron simply because it is rolling around inside of the meson's secondary coils and can't get out. The proton is actually a part of the meson link's secondary coil structure. A proton and meson can be broken apart but they are not separate entities. Cut off a piece of rope and both pieces are still the same rope.

Neutrons are actually formed inside of a meson link during the process of electron capture. When a nuclear isotope has fewer neutrons than it needs to be stable it will use some of its structural energy to collapse one of its electrons into a proton and form a neutron. I do not believe that a neutron can form in any way except through electron capture within a nucleus. A neutron can also be forced inside of a meson during neutron capture and in the process of nuclear fusion.

The neutrons are like rolling balls within an atom's meson. They remain unchanged within the meson until they undergo beta decay. When this happens, the electron is ejected and the proton remains within the nucleus and generates a new magnetic charge chain that eventually couples to the electric charge chain of an external electron. This process is called beta decay.

The alpha particle (He-4 nucleus) is the most stable and has the highest binding energy of any nuclear particle. Its formation takes preference over the formation of any other nuclei. An extreme example of this is the almost instant decay of Beryllium-8. When the structure of a nuclei is disrupted through a decay or a collision, an alpha particle can form spontaneously and become ejected. An alpha particle exists at the center of every nucleus but a nucleus can contain only one alpha particle. Whenever an alpha particle is formed spontaneously from other weak bonds within a nucleus it is immediately ejected.

Fusion is the process by which an alpha particle is formed or a proton is added to a nucleus. Fission is when the bonds between a proton and a neutron come apart. Beta decay results from the splitting of a neutron back into a proton and electron. In the process of beta decay, there is sometimes enough extra energy for the formation of a positron/electron pair. When this happens, the positron soon couples to an electron to form a Positronium atom. Positronium atoms are very well studied low energy combinations of an electron and positron that exist for a short period of time before they annihilate into a number of photons. Positronium exists in the two spin states. Para-positronium has anti-parallel spins and has a lifetime of $(1.25 \times 10^{-10} \text{ sec})$ It decays through the emission of an even number of photons. Ortho-positronium has parallel spins and has a lifetime of $(1.42 \times 10^{-7} \text{ sec})$. It decays through an odd number of photons.

It is through radiation chains that protons and electrons interact with one another and are able to produce photons by combining the last links in both chains. However, for the purpose of understanding the Circlon Model of Nuclear Structure, it is only necessary to consider the first two links in the proton's radiation chain. These are the *proton link* and the *meson link*, which together form the *promestone*. The proton link is identical in size to the meson's secondary coils. These three particles are assembled to form the approximately 2,000 nuclear isotopes of all the elements. These isotopes are constructed according to the Rules Circlon Nuclear Structure.

The Rules of Circlon Nuclear Structure

The following Nuclear Structure Rules describe how Promestones are added, one at a time, to form the nuclear structures of successive elements from Hydrogen (#1) through Circlonium (#118).

To form a stable nucleus, one or more neutrons must be added with each Promestone. As the nucleus grows, one element at a time, its structure must obey the Hydrogen and Alpha Center Rules, and, as structural complexity increases, one or more of the ten other rules.

Hydrogen Rule

Each meson has four Nucleon Receptors equally spaced along its circumference. One of the meson's four Nucleon Receptors must always be occupied by a proton. The other three Nucleon Receptors are spaced at 90 degree intervals from the proton. In the hydrogen nucleus, the Nucleon Receptors at 90 degrees from the proton must remain vacant.

Nucleon Receptors are not physical structures in that they "look" no different from the rest of the meson's circumference; they merely represent the four places where nucleons (protons and neutrons) and other mesons can attach to a meson within a nucleus.

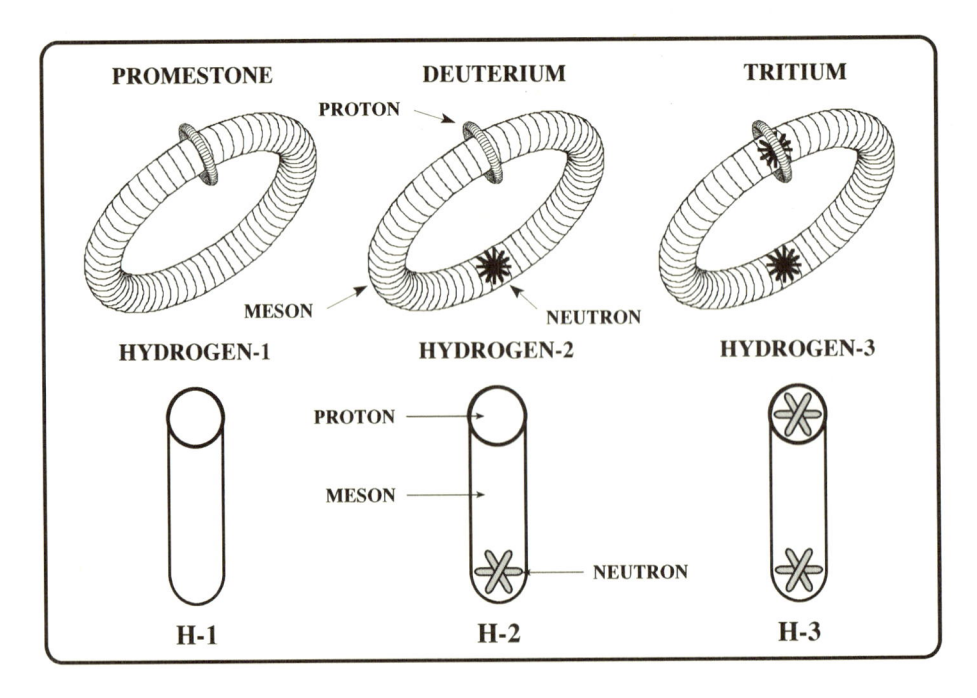

Why Einstein Was an Ignorant Fool — James Carter

The Alpha Center Rule

The center of each nucleus heavier than Hydrogen is formed by an Alpha Center. The structure of the Alpha Center, which is essentially an alpha particle, consists of two mesons crossed at right angles to one another, with a proton and neutron at each intersection.

The two remaining Nucleon Receptors of each meson are vacant so that the He-4 nucleus has four vacant Nucleon Receptors. These four Receptors all contain neutrons in He-8, which is the heaviest unstable isotope of Helium.

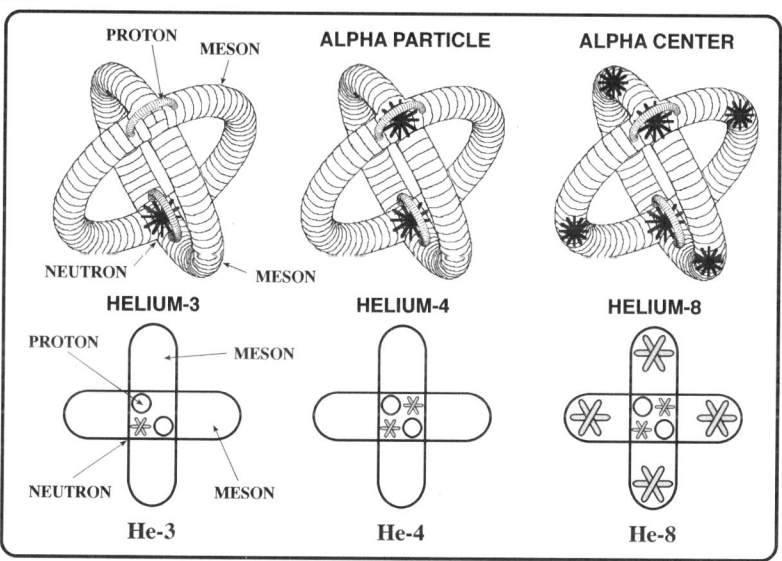

Meson Rules

Four simple rules govern the configuration of protons and neutrons within the mesons that form the completed inner structure of all nuclei large enough for the rules to apply.

Rule of Four The two mesons that form the Alpha Center of a nucleus will each contain four neutrons and four protons when their structure is complete. These two mesons will have one neutron and one proton at each joint where they connect. (This rule applies to all elements from Carbon on.)

Rule of Three All mesons outside of the Alpha Center will contain three neutrons when their structure is complete. (This rule applies to all elements from Sodium on.)

Rule of Two Whenever two mesons are joined together at one point they will contain two nucleons (one neutron and one proton) at this joint when their structure is complete. (This rule applies to all elements from Lithium on.)

Rule of One Whenever two mesons outside of the alpha center are crossed so that they are joined in two places, they will have one proton at one joint and one neutron at the other joint when their structure is complete. (This rule applies to all elements from Nitrogen on.)

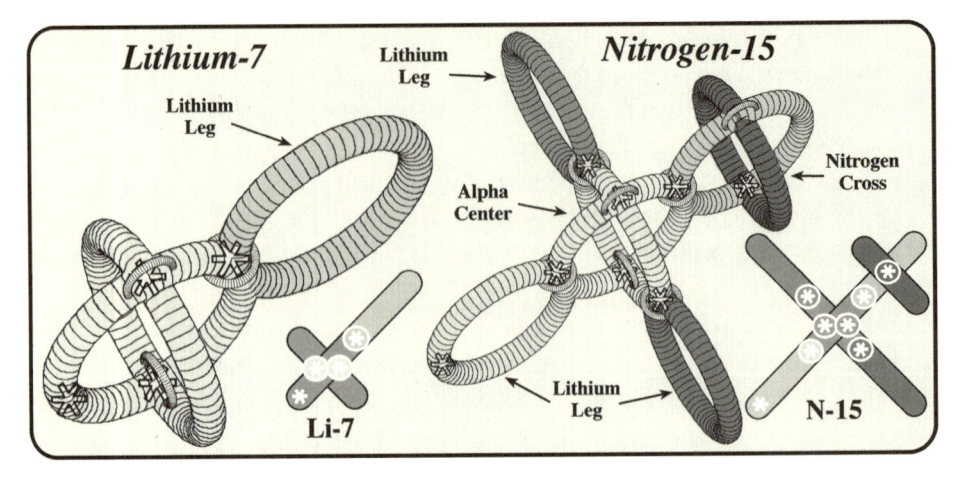

Lithium

Lithium forms when a Promestone attaches to one of the Alpha Center's vacant nucleon receptors. This structure is called a Lithium Leg, and all elements except palladium and the noble gases have at least one. This process is repeated in successive elements, until the Alpha Center's three other vacant receptors are filled with Lithium Legs, forming Carbon.

Nitrogen

Nitrogen forms when a Promestone is attached in a cross formation with one of carbon's four Lithium Legs to form a Nitrogen Cross. Lithium Legs and Nitrogen Crosses hold the electrons of an atom's outermost electron shell.

In a Nitrogen Cross, the proton occupies one pair of crossed nucleon receptors, and the neutron occupies the other pair. The Nitrogen Cross is similar in structure to the Alpha Center, except that its structure is complete when it has one proton at one of the junctions of its crossed mesons, and one neutron at the other junction. This process is repeated with successive elements, until the three remaining Lithium Legs are converted to Nitrogen Crosses to form Neon.

At this point, a second Lithium Process begins with Sodium and ends at Argon to form another outer layer of nuclear structure. This step-by-step building of outer layers of nuclear structure is called the Lithium Process. There are five Lithium Processes, ending with Neon, Argon, Krypton, Xenon, and Radon respectively.

A sixth Lithium Process begins with francium and radium, but is interrupted by the third Scandium Process, and cannot be expected to resume formation until Copernicium #112 and then complete that process at element #118 (Circlonium).

Vanadiun-51   **Pre-Chromium-52**   **Chromium-52**

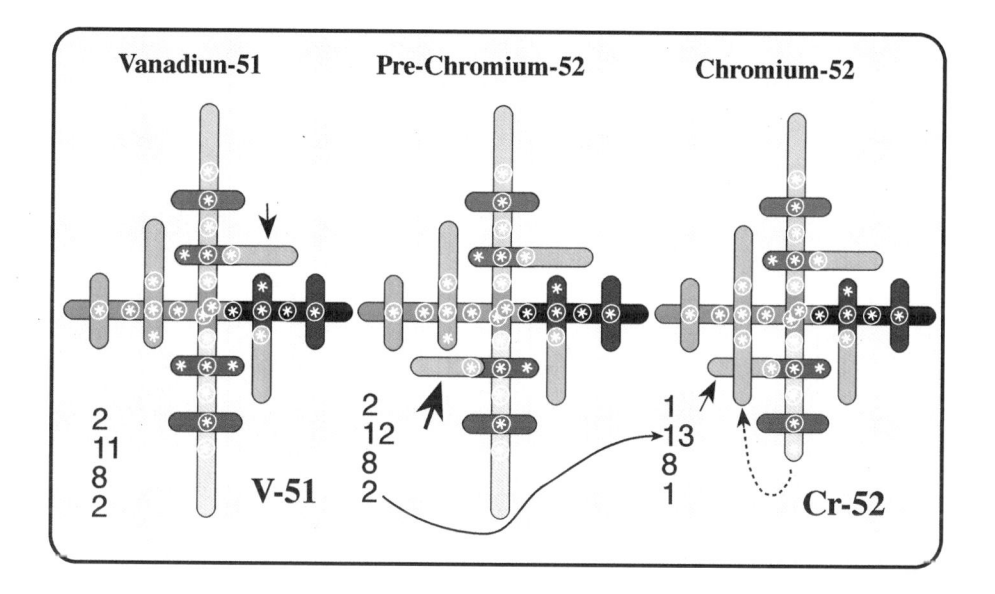

2
11
8
2
   V-51

2
12
8
2

1
13
8
1
   Cr-52

The Dual Event Transformation When a fourth Scandium Ear is added to a Vanadium nucleus, it causes a Promestone from one of its Lithium Legs to immediately move from the third Lithium Layer down into the first Scandium Layer, where it combines with a Scandium Ear to form a Chromium Cross. This is a Dual Event Transformation, and it occurs in the formation of twelve other elements, namely Copper, Niobium, Ruthenium, Palladium, Cerium, Terbium, Gold, Protactinium, Uranium, Neptunium, Plutonium, and Berkelium.

The need for a Dual Event Transformation is indicated in the electron configuration for these elements (see the vertical row of numbers at the lower left of each isotope). These numbers indicate the number of electrons in each of the atom's electron shells. Since each Promestone holds an electron, it shows up in the electron configuration when a Promestone moves from an upper position in the nucleus to a lower one, as the electron held by that Promestone is likewise pulled down into an inner shell.

Dual Event Transformation Rules

Chromium Rule Twenty-five percent of any layer of Chromium Crosses must form in one step and be the result of a Dual Event Transformation. Thus, one Chromium Cross is formed in Chromium and two Chromium Crosses are formed in Cerium and Protactinium.

Niobium Rule When a layer of first four, then three, and finally six Scandium Ears are formed, it immediately initiates a Dual Event Transformation, in which a Promestone moves down into the internal structure of the nucleus from a Lithium Leg.

Why Einstein Was an Ignorant Fool   *James Carter*

This forms a Chromium Cross in the case of chromium, a fourth Scandium Ear in the case of Niobium, and a seventh Scandium Ear in the case of Ruthenium. This rule is not obeyed by elements heavier than Ruthenium.

This is a copy of the Copper block from the
Periodic Table of the Circlon Model of Nuclear Structure

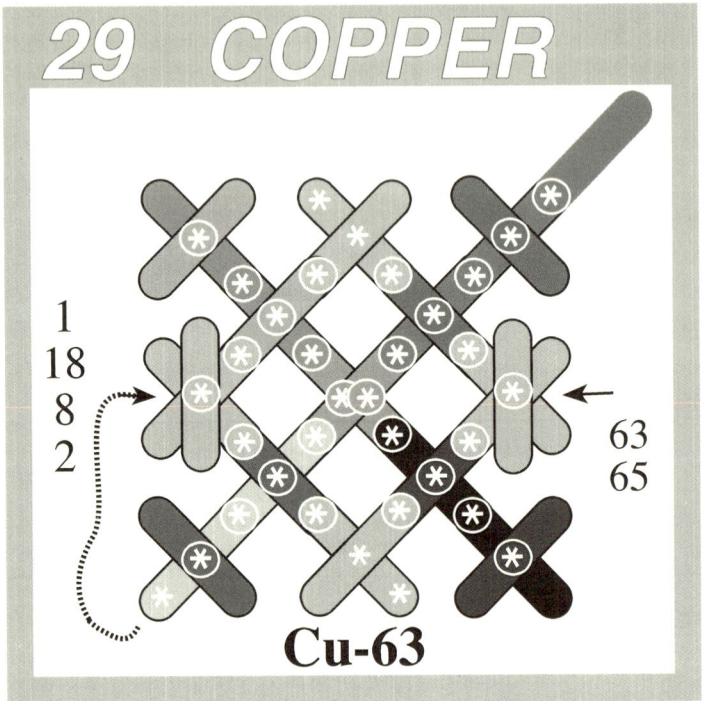

29 COPPER

1
18
8
2

63
65

Cu-63

Copper Balls

Like Chromium, Copper is formed in a Dual Event Transformation, when a Promestone is added to one of Nickel's Chromium Crosses to form a Copper Ball. This creates a dynamical imbalance that causes a Promestone to move down from the Lithium Layer to the Scandium Layer, and form a second Copper Ball opposite the first. In a Copper Ball, the third meson is attached to where the two mesons of the Chromium Cross cross and attach to each other. One of these two junctions contains three mesons and a proton, while the other contains three mesons and a neutron. These two Copper Balls both begin and complete the first layer of two Copper Balls. At this point the third Lithium Process resumes with the addition of a Lithium Leg to form Zinc. Copper's two remaining Chromium Crosses do not become Copper Balls until the formation of Palladium.

Copper Rule

Whenever the last ball in a layer of two, four, or eight Copper Balls is formed, it does so as the result of a Dual Event Transformation, initiated by the formation of either the first, the third, or the seventh ball in the layer. This rule applies to Copper, Palladium, Gold, and Roentgenium.

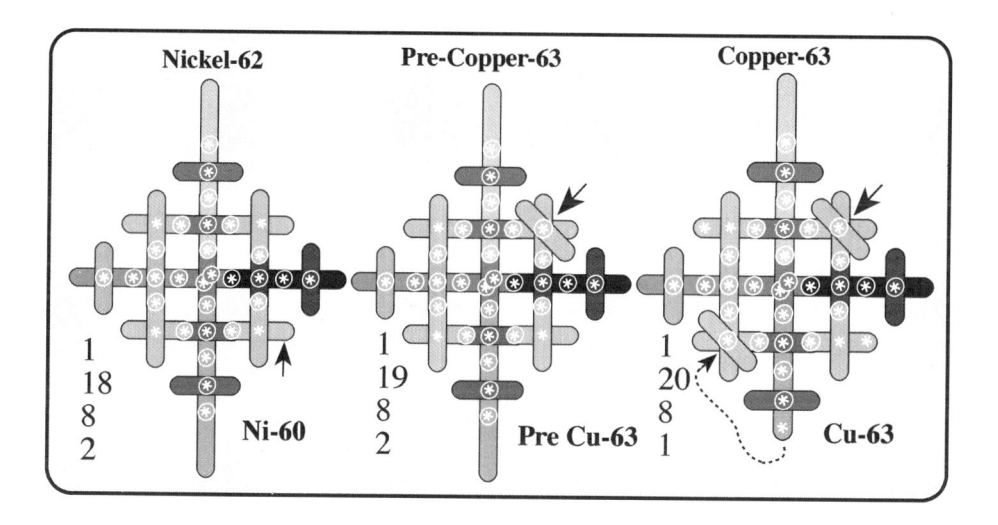

Palladium Rule

The last two balls in a layer of four or eight Copper Balls cannot form until the layer of Scandium Ears of the next Scandium Process has completed its formation. This rule applies to Palladium, Gold, and Roentgenium.

Terbium Rule

Whenever the first ball in a layer of eight Copper Balls is formed, it does so as a result of a Dual Event Transformation, as in the case of Terbium and Berkelium.

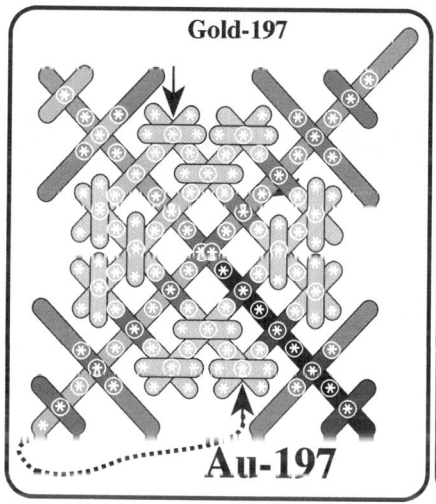

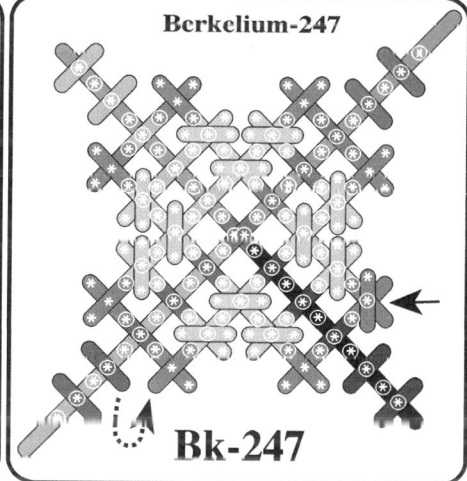

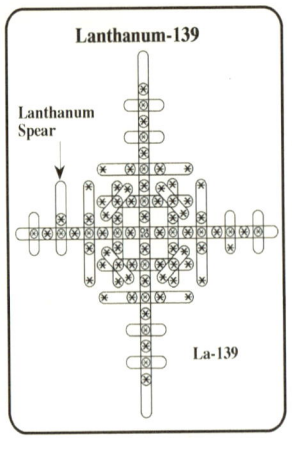

Lanthanum Rule

A Lanthanum Spear will always occur in the element prior to the beginning, and completion, of a layer of eight Chromium Crosses. This rule applies to Lanthanum, Gadolinium, Actinium, and Curium.

    A Lanthanum Spear, which is essentially a false start at the third Scandium Process, is always a temporary nuclear structure that eventually moves down into the internal nuclear structure in a Dual Event Transformation. Lanthanum is formed when a Promestone is attached to one of barium's Nitrogen Legs to form a Lanthanum Spear.

Thorium Rule

In order for the first five Chromium Crosses in the Actinide Group to form, the "pressure" of one Lanthanum Spear must be maintained in the external structure of the nucleus.

    When a Promestone is added to a Thorium nucleus, it forms a third Lanthanum Spear. As soon as this resulting pre-Protactinium nucleus is formed, the two other Lanthanum Spears move down into the third Scandium Layer to form two Chromium Crosses opposite each other. The resulting protactinium nucleus is transformed into Uranium by the addition of another Promestone, which first momentarily forms a second Lanthanum Spear and then either it or the Lanthanum Spear opposite falls down into the third Scandium Layer to form a third Chromium Cross.

    When a Promestone is added to a neptunium nucleus to form a fifth Chromium Cross, the Lanthanum Spear then moves down to form the sixth Chromium Cross of Plutonium.

    This rule applies to Thorium, Protactinium, Uranium, Neptunium. and Plutonium.

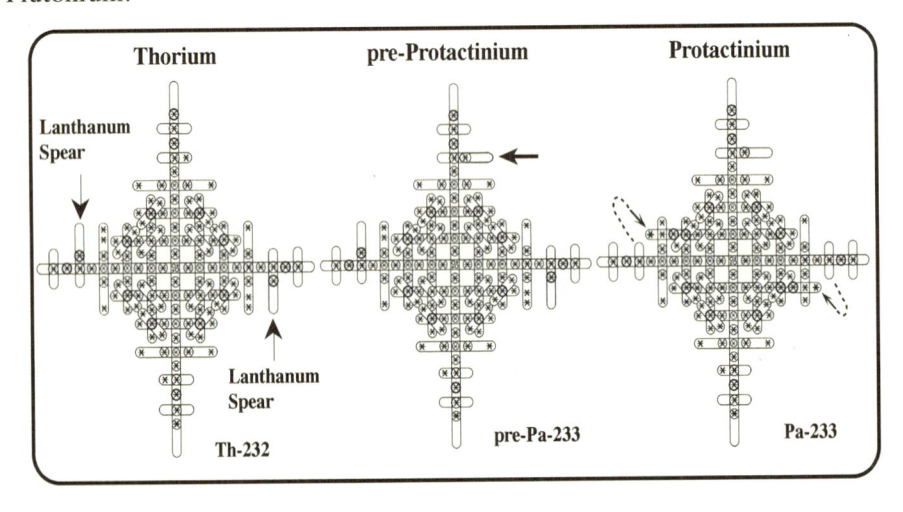

Why Einstein Was an Ignorant Fool   *James Carter*

NUCLEAR GLOSSARY
Archetope Symmetry Principle
All elements have at least 3 known isotopes and some have as many as twenty-nine. The 112 named elements contain nearly two thousand known isotopes. Of these about 280 are stable or have very long lifetimes. Each element has a particular isotope that is most representative of that element. This archetypal isotope is called the element's Archetope. The primary consideration in determining an element's Archetope is symmetry. An Archtope's balance neutrons must maintain an internal symmetry that matches the Archetopes surrounding it on the periodic table. Most Archetopes obey all three Archetope Rules. In almost all cases, an element's Archetope is its most abundant isotope.

Chromium Cross
The nuclear structure that is formed by the crossing and linking together of two adjacent Scandium Ears.

Copper Ball
The nuclear structure that is formed when a Promestone is attached to a Chromium Cross, at a 45 degree angle to that Cross's two component Promestones.

Dual Event Transformation
Dual Event Transformations occur as interruptions in the natural flow of the Scandium Process. When a Dual Event Transformation occurs, the Lithium Process moves one step backwards, enabling the Scandium Process to move two steps forward. They occur at the beginnings and endings of layers and sub-layers in the internal nuclear structure. One explanation of why this occurs, is that the weight of the external nuclear structure is too great for the resistance of the internal nuclear structure, and a Promestone from the external structure "falls" into the internal structure.

Lanthanum Spear
The nuclear structure formed when a Promestone is attached to the side of one of the legs of a nucleus. A Lanthanum Spear is just like a Scandium Ear, except that it attaches farther out on the nuclear leg while it "waits" to migrate down into the internal nuclear structure, to join with a Scandium Ear, to form a Chromium Cross. Lanthanum, Gadolinium, Actinium, Protactinium, Uranium, Neptunium, and Curium each have one Lanthanum Spear, and Thorium is the only element with two.

Lithium Leg
The nuclear structure formed by the attachment of a Promestone to one of helium's vacant nucleon receptors. This process transforms an alpha particle to a lithium nucleus. All elements except the noble gases and Palladium have at least one Lithium Leg in their outer nuclear structure.

Lithium Process
The sequence by which an outer layer of nuclear structure is formed. This is a two-step process, in which first a layer of four Lithium Legs is formed, and then each is transformed into a Nitrogen Cross with the addition of a Promestone.

Meson
The meson is the fourth link in a proton's radiation chain. In the nucleus of the atom, the neutrons fit within the secondary coils of the mesons, and lock the nucleus together.

Nitrogen Cross
The nuclear structure that is formed by the attachment of a Promestone at right angles to the Promestone that forms a Lithium Leg. From Nitrogen on, all elements contain at

Why Einstein Was an Ignorant Fool   *James Carter*

least one Nitrogen Cross in their outer structure, except Silicon, Germanium, Tin, and Lead.

Nucleon Receptors
The four places on the meson where it can attach to protons, neutrons, and other mesons. A proton is always located at one of a meson's Nucleon Receptors and the others are located at 90 degree intervals from the proton.

Promestone
The first two links in the proton's radiation chain. The Promestone is the nucleus of the Hydrogen-1 atom, and as such, along with the neutron, is the basic building block of all the elements. A Promestone with a neutron trapped within the secondary coils of its Meson link is a Deuteron, or Hydrogen-2 nucleus. A Promestone with two neutrons is a Triton, or Hydrogen-3 nucleus.

Scandium Ear
The nuclear structure that is formed when a Promestone is attached to the side of one of the legs of a nucleus and becomes part of its internal nuclear structure.

Scandium Process
The sequence by which an internal layer of nuclear structure is formed. This is a three-step process, in which first a layer of Scandium Ears is formed, which is then converted into a layer of Chromium Crosses, which is then converted into a layer of Copper Balls

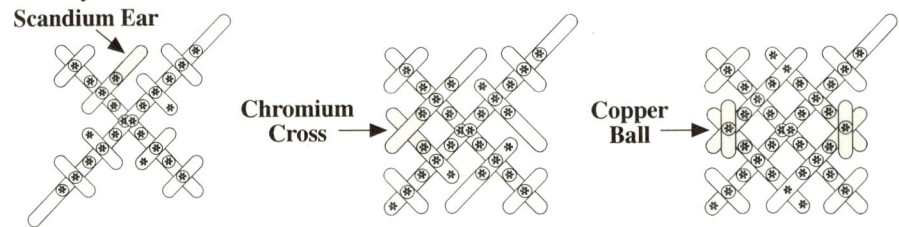

Nuclear Stability Number
The idea of the Nuclear Stability Number is a new concept for the classification of the structure of atomic nuclei. It is a very simple system that matches nuclear structures with a superior degree of accuracy. This system fits the whole range of elements very well. Elements that fall outside of these rules have what are called Stability Anomalies, and provide a means of testing the idea of circlon nuclear structure. These anomalies must be explained in terms of the unique nuclear structure of the elements exhibiting them.

The Stability Number for each element is the increase in mass that its Archetope has over the Archetope of the previous element. This difference in mass is measured in whole units of proton or neutron mass.

For most elements, the Archetope is quite unambiguous, since almost half of the elements have either only one stable isotope or no stable isotopes, and thus only one longest-lived isotope. For most of the other elements with more than one stable isotope, the choice of Archetope is quite straightforward, since the relative natural abundance of the various isotopes of a particular element will usually, quite overwhelmingly, point to a single isotope. These most abundant isotopes almost invariably match the neutron patterns of the Archetopes closely associated with their particular element on the periodic table.

Why Einstein Was an Ignorant Fool   *James Carter*

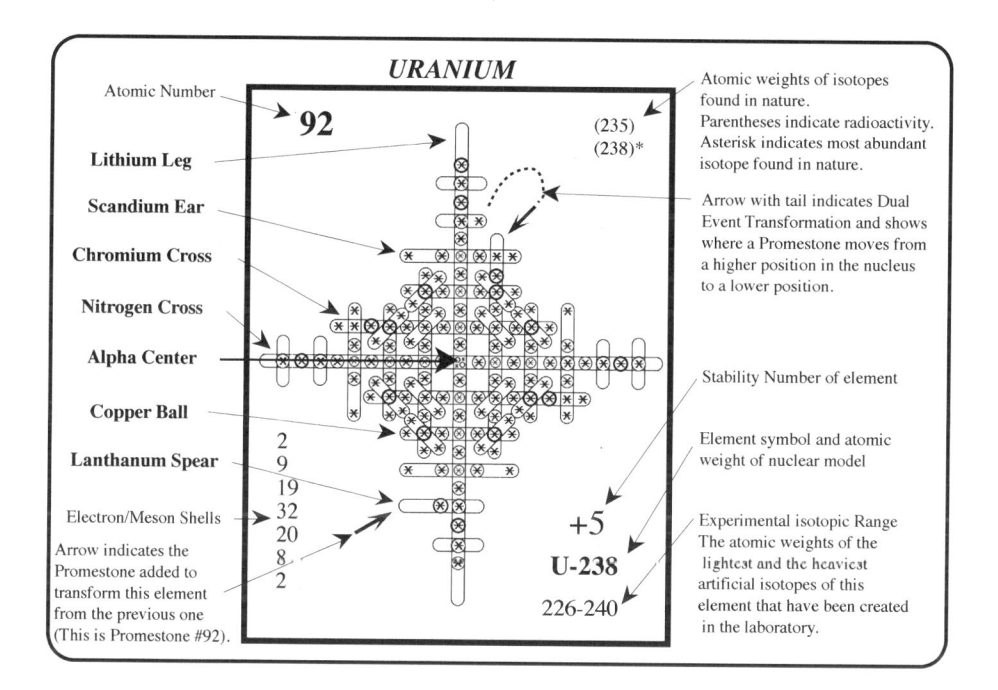

URANIUM

Atomic Number → 92

Lithium Leg
Scandium Ear
Chromium Cross
Nitrogen Cross
Alpha Center
Copper Ball
Lanthanum Spear

Electron/Meson Shells → 2, 9, 19, 32, 20, 8, 2

Arrow indicates the Promestone added to transform this element from the previous one (This is Promestone #92).

(235)
(238)*

+5
U-238
226-240

Atomic weights of isotopes found in nature. Parentheses indicate radioactivity. Asterisk indicates most abundant isotope found in nature.

Arrow with tail indicates Dual Event Transformation and shows where a Promestone moves from a higher position in the nucleus to a lower position.

Stability Number of element

Element symbol and atomic weight of nuclear model

Experimental isotopic Range The atomic weights of the lightest and the heaviest artificial isotopes of this element that have been created in the laboratory.

Archetope Rules

1. The Atomic Weight of Archetopes will be even for even numbered elements and odd for the odd numbered elements.

2. The Archetope of an element is that isotope which is most abundant in nature, or in the case of elements which have no stable isotopes, it is the longest lived isotope.

Number Rules

1. The Stability Number of odd numbered elements to Arsenic is +3, and the Stability Number of odd numbered elements from Arsenic on is +1.

2. The Stability Number of even numbered elements up to Zinc is +1, and the Stability Number of even numbered elements from Zinc on is either +3 or +5.

Stability Anomalies

Any element that violates either the Stability Number Rules or the Archetope Rules has a Stability Anomaly which must be explained in terms of that element's unique nuclear structure, and also in terms of that element's place in the sequential process of nuclear structure.

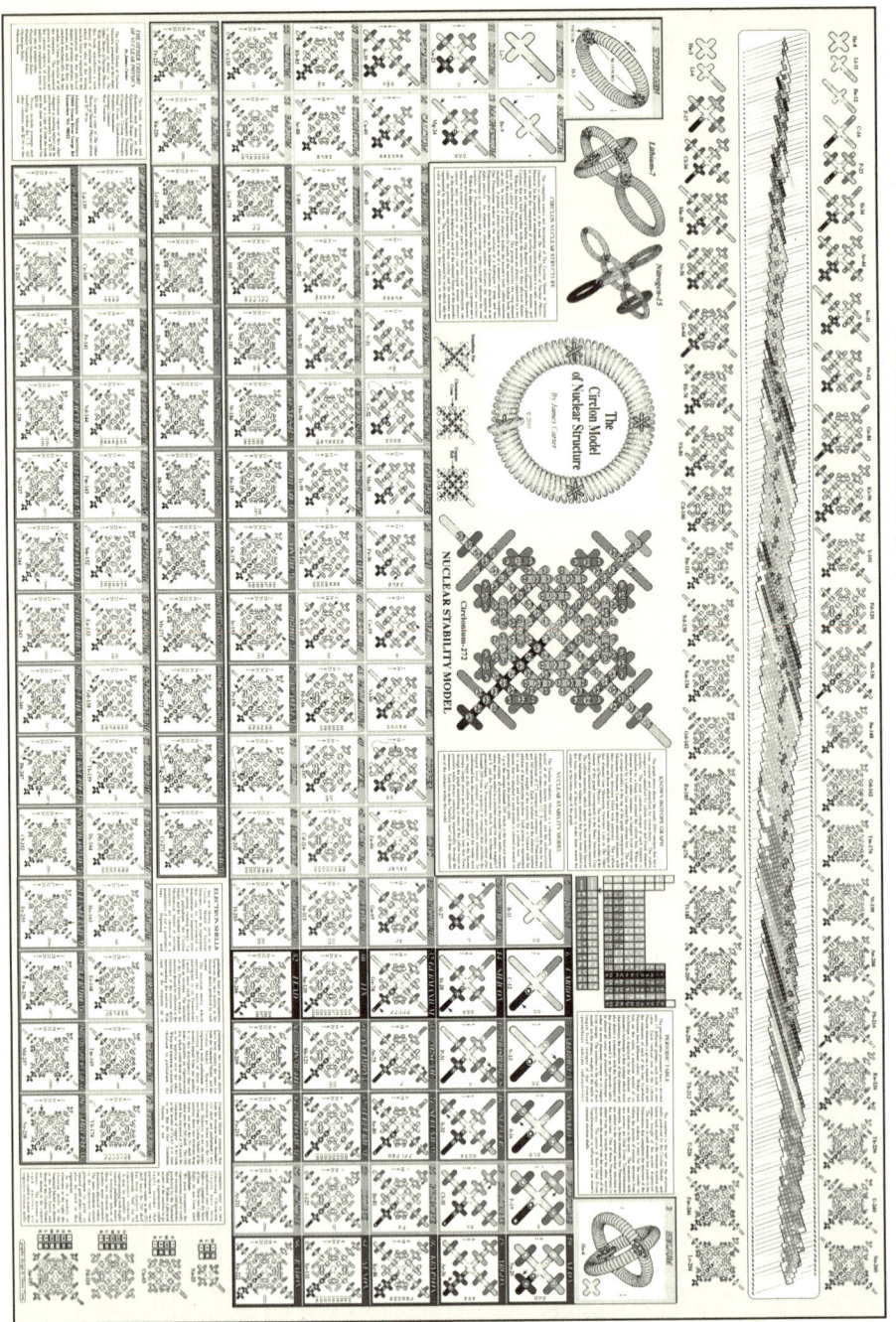

This 39" x 27" full color Circlon Model of Nuclear Structure wall chart contains nuclear models of the most common isotope of each element. My new book "Physics without Metaphysics" also contains these models as well much additional information about the absolute motion of mass, space, time, and gravity and a complete description of Big Bang physics. Both the chart and book can be purchased at www.living-universe.com

Why Einstein Was an Ignorant Fool James Carter

Nuclear Stability Model

The Nuclear Stability Model shown on the back cover is a composite, sequential model of all known elements, up to Copernicium-272. Each is labeled with the element symbol and atomic weight of the nucleus formed with the addition of that particular nucleon. Each proton is represented by a pink circle and each neutron by a yellow box. The meson attached to each proton is colored to match its element group on the periodic table below.

Up to the isotope of Bismuth-209, which is the heaviest known stable isotope, all nucleons in this model form stable isotopes when they are added to a nucleus, except for atomic weights 5 and 8, and the Protons that form Technetium and Promethium. The Protons and neutrons added after Bismuth-209 follow a line through the longest lived isotopes of those elements.

The periodic table presented here is different from a "standard" periodic table in that Lutetium and Lawrencium are moved from the last places in the Lanthanides and Actinides up into the main body of the table. This is because these two elements are each formed with the addition of a Scandium Ear and not with a Chromium Cross or Copper Ball as are the other Lanthanides and Actinide elements.

The periodic table is divided into 13 groups, each with its own color. Each vertical row of the Lithium Process elements is given a separate color, while the elements of each of the four Scandium Processes have different colors.

Silver and Cadmium are shown to be the last two elements of the first Scandium Process even though the last two Copper Balls of this process were formed with palladium. This is to maintain symmetry in both the structure of the model and the coloring system. The Lithium Legs that form Silver and Cadmium first became part of the nuclear structure with the formation of Rubidium and Strontium and then moved down into the internal nuclear structure during the formation of Ruthenium and Palladium. Rather than label these two Lithium Legs twice, it makes more sense to label the last two protons of palladium's Copper Balls as Silver and Cadmium. These elements are thus considered to be the last two members of the first Scandium Process rather than members of the fourth Lithium Process even though they are each formed by the addition of a Lithium Leg. The above explanation is also applicable to Gold and Mercury, and to elements #111 and #112.

NUCLEAR STABILITY MODEL

Copernicium-272

Lithium-7

Why Einstein Was an Ignorant Fool   *James Carter*

Space Aliens and Gravity

To fully understand the principle of the gravitational expansion of mass, space, and time, it is necessary to first present a little science fiction tale. This is a story about the initial discovery of gravitational motion and force by a group of scientists far more advanced than beginners like Galileo, Newton and Einstein.

As our story begins, a group of people are traveling through the deep space between the stars in the Starship Titanic. They have been on this journey for many generations and for one reason or another have lost all knowledge from their former civilization. They are technically advanced and have a wide array of observing and measuring devices. Being far removed from any large bodies of matter for thousands of years, there is no possible reason for any of them to consider the unmeasured phenomenon of gravity. The area of the starship containing their living quarters is a large rotating circular structure that provides an inward centripetal acceleration of 10 m/s^2 at its outer circular floor. This artificial gravity saves their bodies from the damaging effects of being weightless for long periods of time. These space travelers understand this constant centripetal acceleration very well and are able to increase or decrease it by changing the rate of rotation or moving up or down within the capsule. They are also familiar with the momentary linear acceleration produced by the starship's engines. It is just that they do not associate these accelerations with gravity because they have neither heard of nor experienced the phenomenon of gravity.

At long last, the Titanic approaches a star surrounded by a group of planets similar to our own solar system. After decelerating their spacecraft to the inertial frame of the star, the space aliens decide to stop and explore one of the inner planets that has oceans and continents. They were looking for a new planet to call home and this looked like a likely candidate.

As they maneuvered their spaceship around the planet, they unexpectedly noticed that its surface seemed to rush towards them when they were not measuring any acceleration toward it. After a number of speed adjustments and calculations they determined that they could set their ship at a velocity vector that was both away from the planet and at right angles to it and that exactly balanced the motion of the planet's surface toward them. These maneuvers put them in a circular orbit around the planet. Eventually, they guided the ship to a stationary point above the rotating planet's equator (geosynchronous orbit).

They didn't quite understand the actual dynamics of this orbit, but since the Titanic seemed to be secure as it revolved with the planet, they decided to send a smaller exploratory craft down to the planet's surface where they could observe this strange new phenomenon from close up.

As they moved down toward the planet's surface, they had to keep accelerating the craft upward in order to eventually make a soft landing. Once they arrived on the surface, they were able to measure that it was accelerating upward at $10m/s^2$ just like it had been appearing to do from orbit. Once the explorers got out of the craft and started walking around, they quickly noticed that the upward acceleration of the ground felt exactly like the inward centripetal force of their rotating circular living quarters back on the Titanic.

After thinking about this strange phenomenon for some time, they concluded that the matter within the planet must be slowly expanding. This idea would also explain how the Titanic was able to maintain its orbit around the planet. It was simply moving away from the planet at the same speed that the planet's surface was moving toward it. While the idea of an expanding planet seemed rather strange and unexpected, that was certainly what they had measured with their accelerometers and almost no one was able to come up with an alternative explanation.

One old man with fuzzy white hair suggested the somewhat incomprehensible theory that there was some kind of a virtual attraction between the people and the planet that constantly pulled them together. This idea seemed completely out of the question to the majority of the observers because such an unmeasured effect would be completely unlike the effects of magnetic and electrical attractions that they were all used to calculating and measuring. The concept of an infinite reaching gravitational attraction between individual atoms also seemed to be completely unreasonable because the simpler and more intuitive local measurement of gravitational expansion was so readily at hand.

These were practical people who were used to the everyday measuring of force and acceleration in a straightforward way. The idea of an infinite and inalterable attraction between all bodies of matter and photons was so preposterous that such an unlikely and counter-intuitive concept could simply not be accepted by any of them.

They concluded that if the matter within the planet was slowly expanding, then the matter contained within the Titanic and even their physical bodies was also expending at an, until now, imperceptible rate. To test their new principle of gravitational expansion they placed sensitive accelerometers at both the Titanic's bow and stern. They discovered that the two ends of the Titanic were accelerating and moving away from each other at a small but measurable velocity. The old man with the white hair claimed their experiment proved nothing since his mass attraction theory would produce the same results.

The technicians got together and designed an experiment in which gravitational expansion and gravitational attraction would yield different results. They fashioned a large glass sphere with a hollow shaft through its center. The then machined a solid Gold ball that would loosely fit within the shaft. They attached a number of accelerometers to the sphere's surface and put another one at the Gold ball's center. They then placed the ball at rest in the hole at the outer surface of the sphere and then recorded video of the apparatus as the ball began to move down the shaft towards the center of the glass sphere.

The old man predicted the ball would be attracted toward the center of mass and accelerate to a maximum velocity at the center and then decelerate to a stop at the opposite end of the hole. He claimed this back and forth motion would repeat endlessly barring any friction between the ball and the sphere. The rest of the group predicted that gravitational expansion would leave the ball motionless while the surface of the sphere moved away from it in all directions.

As they watched, the ball appeared to decelerate to stop at the center. When they checked the accelerometers they found that all points on the sphere's surface continued to accelerate away from its center while the ball registered no acceleration at all. They concluded that gravitation attraction could not exist because, like Einstein, they were unable to devise any experiment that could detect it.

In conclusion, no rational space alien who carefully measured gravity would ever conclude that it is some kind of occult and non-local attraction between atoms when it can be easily measured that the mass, space, and time of all atoms are slowing expanding at a constant and synchronous rate throughout the Living-Universe.

The Gravity Cannon Test
*The gravity cannon is a definitive experimental test that can easily differentiate between the four possible general theories of gravity. This test is so simple and basic that once it has been performed, the results can be put on **You Tube** to make it possible for the true nature of gravity to be understood by everyone. Even the small child will be able to clearly see and understand just how gravity really works.*

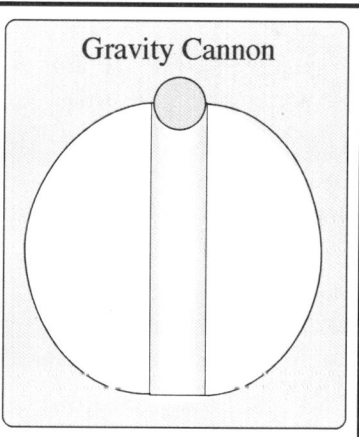

A Definitive Test for the Law of Gravity
The gravity cannon experiment is a very simple test that can provide decisive confirmation or falsification of the various gravity theories.

Although it would have to be performed in the weightlessness of outer space, this experiment could otherwise be executed very easily and inexpensively. A Gold ball would loosely fit inside of a hole passing through the center of a solid, clear glass sphere. When a motionless ball is placed at the barrel's mouth, the gravitational force and motion that occurs is measured and recorded with a video camera.

Besides being able to differentiate between the four possible theories of gravitational force and motion, this experiment would also provides the means to verify and calibrate earthbound measurements and calculations of both the Newtonian force constant G and the gravitational velocity constant G_v. This is a definitive test to identify the physical differences between the four theories of gravity. This test measures the speed of gravitational motion without the influence of any other gravitating bodies or electric and magnetic fields and compares these values to the predictions of gravity theories.

The Four Gravity Theories

There are four different basic gravity theories. These include the two pulling medium theories of homogeneous and infinite particle aether. The other two are the internal and external pushing particle theories.

These four ideas each use different complementary equations to explain the many measurements that have been made of gravitational force and motion. Depending on their various assumptions, all these theories can be made to account for most gravity measurements.

The Gravity Cannon Experiment is based on the Newtonian momentum measurement principle of $F = ma$. This principle shows that the direction of gravitational force and motion
points up at Earth's surface and produces an upward acceleration of about 10m/s^2 and an upward escape/surface velocity of about 11 km/s. Newtonian force and motion is not a theory. It is the principle of measurement that theories of gravity attempt to explain. Both the aether theories and the external pushing particle theories base their ideas on the unmeasurable premise of a metaphysical gravitational force that extends to infinity but points downward toward the centers of all bodies. This force is calculated to be equivalent to and opposite of the upward force of gravity that is measured as Newtonian acceleration.

Aether Gravity Theories

Aethers and fields are defined as any description or condition of space that is not an eternal dimensionless void. Fields are local conditions of aether or spacetime that extend between and connect atoms. They can be either local to the atoms or they can extend to infinity.

The homogeneous aether theories explain gravity as a single, universal, solid or liquid all pervasive aether continuum. Curvatures, ripples, and waves within this universal substance cause bodies of matter to move toward one another. In the infinite particle aether theories, gravity is explained by a potentially infinite number of gravitational particles, waves, or fields that are usually called gravitons. These calculated wave-particle dualities are generated at the center of each body of mass and then spread out in all directions to infinity at the speed of light.

General Relativity is a homogeneous aether theory that has been mathematically crafted into several interpretations. Its equations usually calculate a four-dimensional spacetime continuum that connects all matter and interacts with an apparent but otherwise undetectable force that causes gravitational force and motion. The presence of a body of mass causes the continuum to curve and produce motion in the body. General Relativity is sometimes classed as an infinite particle aether theory because in some versions the force of gravity is spread from atom to atom across the universe by great numbers of tiny wave-particle dualities called gravitons. These wavelike particles move through the continuum at the speed of light and are calculated to cause portions of the spacetime to curve in such a way as to cause the appearance of gravitational motion between bodies of matter.

Pushing Particle Gravity Theories

The pushing particle theories are divided into the external theories and the internal principle of the gravitational expansion of mass, space, and time. The external pushing gravity theories explain gravity by assuming that large bodies of matter like Earth are constantly being pushed inward toward their centers by great numbers of tiny undetectable particles impinging on them from all directions of space. The internal pushing particle principle explains gravity as the measured outward force caused by protons and electrons pushing on one another.

External pushing gravity theories claim that the imagined downward motion of falling bodies is produced by the absorption of tiny undetectable extremely high speed particles that are assumed to exist uniformly distributed throughout all of space. Some of these theories predict particle speeds many orders of magnitude greater than the speed of light. When these particles strike matter, they give it a slight push. These omnidirectional particles push the surfaces of large bodies towards their centers. Such theories predict that the Gold ball would be pushed back and forth from one side of the glass sphere to the other in a similar manner to the predictions of pulling gravity aether theories.

An external pushing gravity theory was first proposed by Nicolas Fatio in 1690. Later, similar theories were proposed by Le Sage and others. Rene Descartes had a pushing gravity theory in which numerous tiny whirlpools within the aether pushed on matter.

While external pushing gravity theories have never gained much credibility among the physics establishment, they have a wide following among alternative gravitational theorists. These theories have no explanation for the equivalence principle and generally ignore the concept altogether.

In the internal pushing particle principle, the particles that do the pushing are the well established protons and electrons within atoms. This explanation of gravity is a principle of measurement and not a theory because the outward force of gravity is easily measured with accelerometers. The motion of the gravity cannonball can easily indicate the truth between the external and internal pushing particle explanations of gravity. In the internal pushing particle principle of the proton and electron, the Gold ball would move from the surface of the sphere to its center where it would gradually slow to a stop. All of the pushing forces are contained within the glass sphere and the Gold ball and there is no force between them.

The 3 Possibilities for Gravity

Gravity can only be a downward pull, a downward push or an upward push. Almost all theoretical physicists imagine it to be a downward pull and a few believe it to be a downward push but only experimental physicists know it to be a measured upward push. This experiment will measure only the gravitational motion between two spheres that do not touch.

There can be only two possible outcomes to the gravity cannon experiment. This test will decide the absolute physical truth between whether gravity points down with equivalent acceleration as Einstein imagined and calculated or whether the force of gravity points up as has always been felt and measured. Either the Gold ball will appear to move to the center of mass and stop or it will be pulled or pushed back and forth from one end of the barrel to the other. The test will determine once and for all whether Einstein's equivalent force and motion are real or whether only measured inertial momentum and force are real. Einstein believed that his equivalent force could add to and then subtract equivalent momentum from the Gold ball as it moved through the glass barrel.

The internal proton/electron pushing principle of absolute gravitational force and motion predicts that the ball will appear to begin accelerating toward the center and then slow to a stop with decreasing deceleration at the sphere's center. The ball remains at rest while the outer surfaces of both ball and sphere move away from their inertial centers. There is no absolute motion between the inertial centers of ball and sphere.

The Orbiting Chain

To demonstrate an orbit around the Earth, we will first describe an experiment that was available even in Galileo's time. A powerful cannon is fired over the surface of the Earth and the path of the cannonball is recorded. The cannon is then again fired from the point where the first cannonball struck. This process continues until the cannonball has traveled all the way around the Earth.

Orbiting Chains

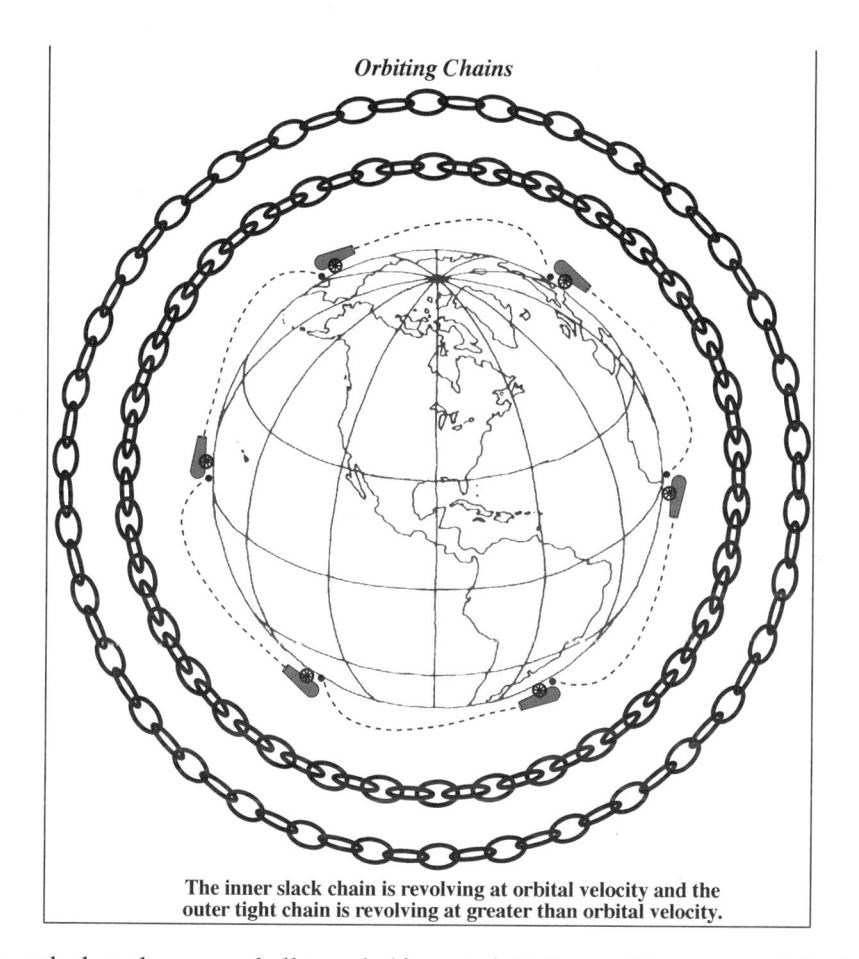

The inner slack chain is revolving at orbital velocity and the outer tight chain is revolving at greater than orbital velocity.

In each shot, the cannonball traveled in a straight line until it was struck by the upwardly moving Earth. However, any photos of the cannonball's path would show it to have followed a parabolic curve. In this digital orbit of the Earth, the cannonball always travels in a straight inertial line but at the same time its path always seems to curve downward. This apparent non-inertial curvature of the Earth's internal space results from the expanding dimensions of matter.

The orbiting chain is another possible model for creating an orbit around the expanding Earth. The chain is wrapped around the Earth and then spun at a high velocity. As the chain goes faster and faster it tightens up and goes into an Earth orbit defined by its length. The faster the chain is spun beyond its orbital velocity, the tighter it becomes due to its increasing centripetal force.

To better understand how orbits work around gravitationally expanding bodies of matter, we can cause the chain to slow until its centripetal acceleration becomes less than the acceleration of gravity. The individual links slacken and lose their tension with one another. However, the slack chain as a whole still maintains its overall orbit while each loosely connected link maintains its own individual orbit without physically touching other links. The dynamics of this orbiting chain satellite are the same whether we use the mechanics of

Why Einstein Was an Ignorant Fool   *James Carter*

gravitational expansion or the gravitational field theory of Newton or the curved spacetime of Einstein.

Principle of the Gravitational Expansion of Mass, Space, and Time
Absolute gravitational motion and force is not a theory of gravity. It is just the measurement of gravity that reveals why we have always felt the upward push of Earth's surface. Chicken little was wrong. The sky is not falling. Earth is falling up!

The principle of gravitational expansion reveals gravity totally in terms of its physical measurements with no metaphysical assumptions such as aethers, fields, actions at a distance, or unseen impinging particles from space. Expanding mass, space, and time show that our measurements of gravitational force are real and that the acceleration of gravity produces true upward motion. Gravity is merely the outward force produced by the gradual and constant dimensional expansion of mass, space, and time. A falling body does not accelerate downward because no such change in motion can be measured. Like the Gold ball in the gravity cannon, falling bodies do not change their state of motion while the surface of Earth moves upward with measured Newtonian acceleration and velocity. Gravity and inertia are not just equivalent. They are exactly equal because they are the same thing.

Successful Performance of the Gravity Cannon Experiment Could Save Billions of Dollars for World Governments.

The gravity cannon test will provide a decisive experimental and mathematical difference between both aether theories and the many pushing gravity particle theories. In particular, it will demonstrate a precise difference between the principle of the gravitational expansion of mass, space, and time and the theory of general relativity as well as all other theories of gravitation.

Other previous precise experimental measurements of gravity such as GPS clock rates and the Pound-Rebka shifts tended to yield the same predicted results for both gravitational expansion and general relativity. The beauty of the Gravity Cannon Experiment is that the principle of gravitational expansion predicts a cause and effect event that is opposite to the effects of all the other gravity theories' causes. General relativity and other gravity theories may predict similar results but no theory predicts the same values as the principle of gravitational measurement.

The ultimate benefits of putting a gravity cannon in orbit could be enormous. Hundreds of millions of dollars have been spent to test one or another of General Relativity's many predictions. Just one example is the LIGO experiment that is attempting to measure Einstein's predicted continuum of gravitons and gravity waves. The gravity cannon would either verify General Relativity's curved space interpretation of gravity or prove the opposite curving matter interpretation of absolute gravitational expansion. If this experiment verifies gravitational expansion, governmental scientific organizations can save hundreds of millions of dollars by not testing Einstein's foolish theories. *Once and for all, we will be able to know with certainty whether gravity points up or down and whether it is a push or a pull.*

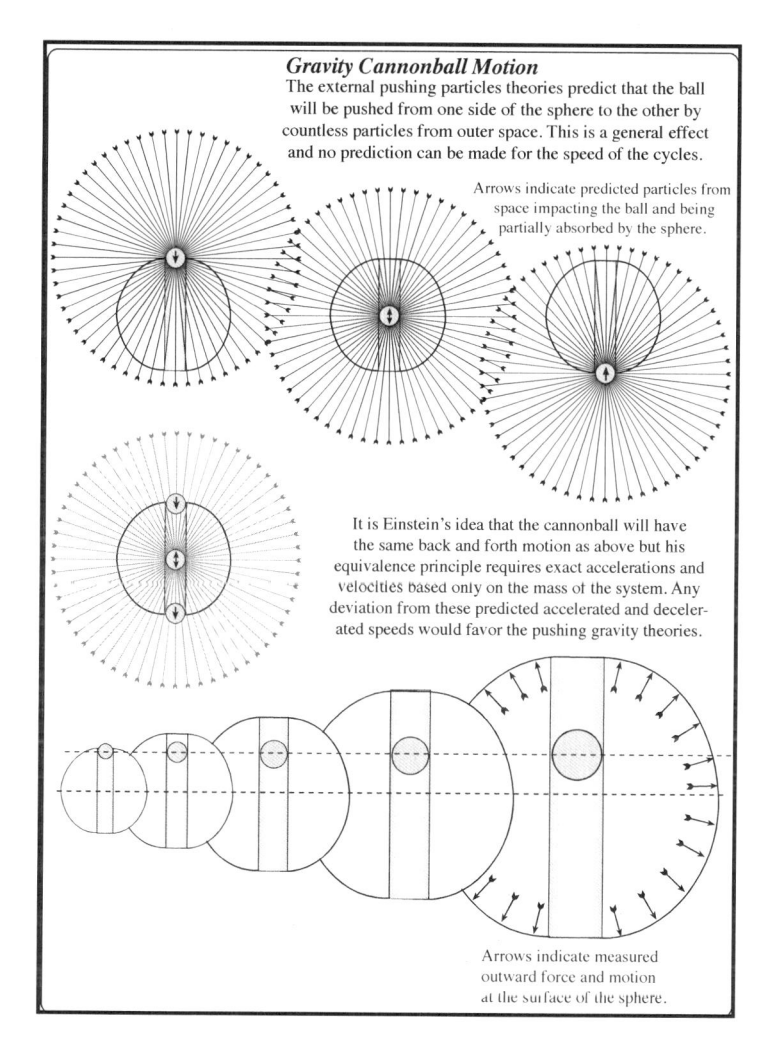

Gravity Cannonball Motion
The external pushing particles theories predict that the ball will be pushed from one side of the sphere to the other by countless particles from outer space. This is a general effect and no prediction can be made for the speed of the cycles.

Arrows indicate predicted particles from space impacting the ball and being partially absorbed by the sphere.

It is Einstein's idea that the cannonball will have the same back and forth motion as above but his equivalence principle requires exact accelerations and velocities based only on the mass of the system. Any deviation from these predicted accelerated and decelerated speeds would favor the pushing gravity theories.

Arrows indicate measured outward force and motion at the surface of the sphere.

Gravity Cannon Experiment
If gravity is not the infinite pulling and curving of spacetime throughout the universe then maybe gravity's quantum nature is just the purely local mechanical event of one atom pushing against another.

The principle of the gravitational expansion of mass, space, and time measures outward acceleration at the surface of the glass sphere, but no acceleration can be measured of the ball toward the centers of mass. Without physical acceleration, the ball will appear to move toward the center of the sphere with decreasing deceleration but will not have the inertial motion necessary to move past the mass center.

Einstein's idea of equivalent gravity predicts the opposite outcome for the gravity cannon experiment. Newton, Einstein, and all other gravitation theorists make the complex prediction that the cannonball will accelerate toward the center of the sphere at a decreasing rate. It will reach maximum velocity at center and then is predicted to decelerate to a stop at the end of the barrel. This starts a new cycle where the ball accelerates back to the center and then decelerates to the other end of the barrel.

Joules of the Living Universe

This diagram shows the quantity of energy measured in Joules for everything from the extremely week photons from AC transmission lines to the total energy of the Big Bang and the creation of the universe.

    This diagram shows the energy transfer of a number of familiar events over the whole energy spectrum. Some of the least energetic photons that we commonly measure are those produced by 60 cycle **alternating current power lines**. At 18,000 km their wavelengths are almost three times greater than the Earth's diameter and 10,000 times more energetic are **AM radio photons**, which have wavelengths of just less than one kilometer. **Television photons** have wavelengths of a few meters and **2.7° CBR photons** are most intense at a wavelength of about one-millimeter. Several thousand times smaller are the many differently colored photons of the **visible spectrum**. Another million times smaller and more energetic are the photons produced when an **electron and positron** join and transform into photons.

Why Einstein Was an Ignorant Fool   *James Carter*

Up to this point on the scale, all individual energy units had been photons. The energy of the next item, **hydrogen fusion**, is released not as photons but as the kinetic energy of the Helium and Hydrogen nuclei and neutrons speeding away from one another. A few times more energy is produced in the fission of an atom of **Uranium-235**. Photons are produced in these events, but the majority of the energy is contained in rapidly moving neutrons and nuclei. The two photons produced in **proton-antiproton annihilation** each have energies of just under one billion electron volts.

The next few items on the scale are the units of energy used in science and commerce. The **erg** is the energy of a gram of mass moving at a velocity of $\sqrt{2}$ centimeters per second. A **foot pound** is the energy of one pound moving at $\sqrt{2}$ feet per second. A **calorie** is the amount of energy required to raise the temperature of one gram of water one degree Celsius.

At about 50 Joules, the **most energetic cosmic rays** to be measured are probably photons with wavelengths of about 10^{-25} meters and masses of about 10^{-15} kilograms. They are part of a background photon spectrum that originated when the matter of the universe was created. The photons at the upper end of this creation spectrum are extremely rare but they have almost unlimited energies with masses approaching that of the universe itself. The masses and wavelengths of selected photons from this spectrum are shown at random points on the spiral.

One **British thermal unit** raises the temperature of a pound of water by one degree. A **watt-hour** is a standard unit of electricity. A **horsepower** is a unit of work. **Natural gas, gasoline, TNT, coal, and hydrogen** are common fuels used to produce energy.

The kinetic energy of a **human being's motion relative to the CBR photon rest** is similar to the energies of the **Titanic's fall** to the bottom of the ocean, the **first atomic bomb** exploded in New Mexico or a **bumblebee** flying at half the speed of light.

The photon energy produced by the **annihilation of one gram of matter with one gram of antimatter** is similar to the energy produced by the **fission of 1000 grams of uranium-235**

The mysterious explosion that occurred in 1908 near the **Tunguska** River in Siberia could well have been caused by a photon from the creation photon spectrum hitting the atmosphere.

The kinetic energy of a **human being moving at half the speed of light** is only a few times less than the **annual electricity production of the United States**. The **Earth's daily receipt of energy** from the sun is about 10 times greater than all the **electricity generated since Tesla** both invented and discovered three-phase alternating current and about 30 times less than the kinetic energy of the **Titanic moving at half the speed of light.**

The kinetic energy inherent in the **earth's motion relative to CBR photon rest** is equal to **the sun's total energy output** for about 100 years. If the **earth were moving at half the speed of light** its kinetic energy would be equal to the sun's output for about 10,000,000 years.

On **May 8th 1997 a gamma ray burst** was detected far off in the universe that released as much energy in a few seconds as the sun has produced since its formation. On **December 14th** of that year a much larger burst was detected that produced more energy than the entire Milky Way galaxy puts out in 10,000 years and even larger bursts have been measured since. Gamma ray bursts are most likely to be caused by photons from the upper end of the creation photon spectrum hitting bodies of matter in the universe at large.

Physics without Metaphysics
By James Carter

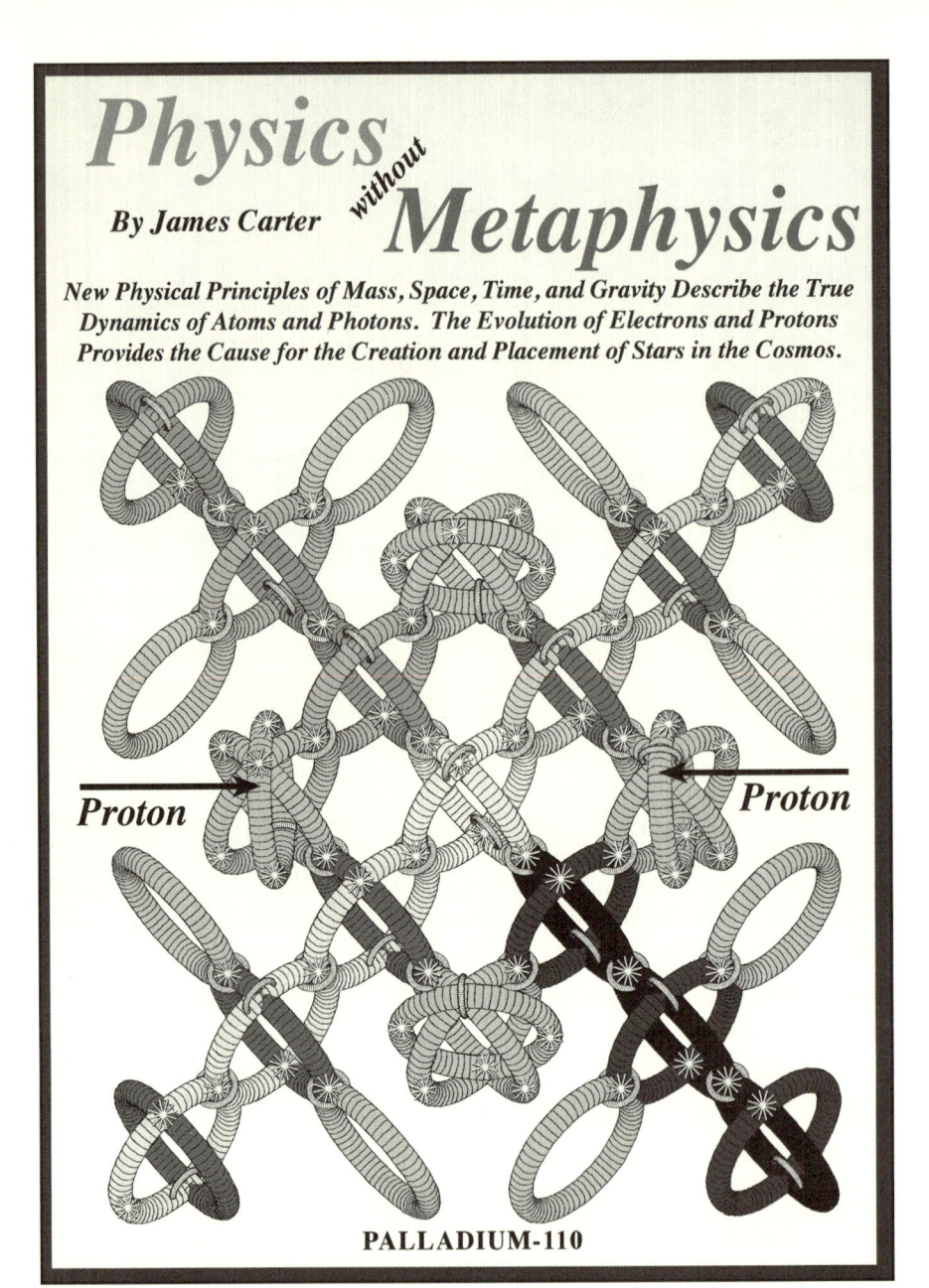

New Physical Principles of Mass, Space, Time, and Gravity Describe the True Dynamics of Atoms and Photons. The Evolution of Electrons and Protons Provides the Cause for the Creation and Placement of Stars in the Cosmos.

Proton ← → *Proton*

PALLADIUM-110

Physics Without Metaphysics **contains a complete description of the Living Universe's cosmic creation process. It illustrates full color models of all nuclear isotopes with circlon shapes for electrons, protons, neutrons, and photons. The book contains many illustrations of gravity and photon experiments and explanations of astronomical observations. To purchase this, or other books by James Carter go to: www.living-universe.com. The site contains more information on Circlon Synchronicity, the Big Bang, gravitational expansion and explains Plate Tectonics and large dinosaur bones.**